The Bering Sea

ALASKA GEOGRAPHIC® Volume 26, Number 3 / 1999

To teach many more to better know and more wisely use our natural resources

COLOR SEPARATIONS: Graphic Chromatics

PRINTING: Hart Press

ISBN: 1-56661-048-6

PRICE TO NON-MEMBERS THIS ISSUE:
$21.95

POSTMASTER:
Send address changes to:

ALASKA GEOGRAPHIC®
P.O. Box 93370, Anchorage, Alaska 99509-3370

PRINTED IN U.S.A.

COVER: *Red-legged kittiwakes are endemic to the Bering Sea region. (Gary Schultz)*

PREVIOUS PAGE: *The Bering Sea supports the Aleut residents of St. George, population 173, in the Pribilof Islands. (Fred Hirschmann)*

FACING PAGE: *A fisherman unloads his catch of sockeye salmon at Togiak, population 801. (Fred Hirschmann)*

ABOUT THIS ISSUE:

We are grateful to the many scientists and researchers who provided information and reviewed portions of this manuscript. Special thanks are due Allen Macklin and Phyllis Stabeno of the National Oceanic and Atmospheric Administration's Pacific Marine Environmental Laboratory, in Seattle; Vera Alexander, Don Schell and Tom Weingartner, all of the University of Alaska's Institute of Marine Science, in Fairbanks; and Andrei Kurbatov, of the State University of New York in Buffalo. We are indebted also to John Piatt, seabird specialist with U.S. Geological Survey's Biological Resources Division, in Anchorage; to Connie Barclay, Bob Platte, Rosa Meehan, Susanne Kalxdorff and Linda Comerci of the U.S. Fish and Wildlife Service; to Kathy Frost and Lloyd Lowry of the Alaska Department of Fish and Game and to Dorothy Childers and her staff at the Alaska Marine Conservation Council, in Anchorage.

Community populations noted in this issue are the latest figures available from the Alaska Department of Commerce and Rural Development.

The Library of Congress has cataloged this serial publication as follows:

Alaska Geographic. v.1-
[Anchorage, Alaska Geographic Society] 1972-
v. ill. (part col.). 23 x 31 cm.
Quarterly
Official publication of The Alaska Geographic Society.
Key title: Alaska geographic, ISSN 0361-1353.

1. Alaska—Description and travel—1959-
—Periodicals. I. Alaska Geographic Society.

F901.A266 917.98'04'505 72-92087

Library of Congress 75[79112] MARC-S.

Robert A. Henning
1915 –1999

We are sad to report the death of Bob Henning on June 6, 1999, in Seattle. A passionate champion of Alaska and of geography, Bob founded The Alaska Geographic Society in 1968. We will miss his enthusiasm and his ideas.

ALASKA GEOGRAPHIC® (ISSN 0361-1353) is published quarterly by The Alaska Geographic Society, 639 West International Airport Rd., Unit 38, Anchorage, AK 99518. Periodicals postage paid at Anchorage, Alaska, and additional mailing offices. Registered trademark: Alaska Geographic, ISSN 0361-1353; key title Alaska Geographic. This issue published Sept. 1999.

THE ALASKA GEOGRAPHIC SOCIETY is a non-profit, educational organization dedicated to improving geographic understanding of Alaska and the North, putting geography back in the classroom and exploring new methods of teaching and learning.

MEMBERS RECEIVE *ALASKA GEOGRAPHIC®*, a high-quality, colorful quar-terly that devotes each issue to monographic, in-depth coverage of a specific northern region or resource-oriented subject. Back issues are also available (see p. 96). Membership is $49 ($59 to non-U.S. addresses) per year. To order or to request a free catalog of back issues, contact: Alaska Geographic Society, P.O. Box 93370, Anchorage, AK 99509-3370; phone (907) 562-0164 or toll free (888) 255-6697, fax (907) 562-0479, e-mail: akgeo@akgeo.com, web: www.akgeo.com

SUBMITTING PHOTOGRAPHS: Those interested in submitting photos for possible publication should write for a list of upcoming topics or other photo needs and a copy of our editorial guidelines. We cannot be responsible for unsolicited submissions. Submissions not accompanied by sufficient postage for return by certified mail will be returned by regular mail.

CHANGE OF ADDRESS: The post office will not automatically forward *ALASKA GEOGRAPHIC®* when you move. To ensure continuous service, please notify us at least six weeks before moving. Send your new address and membership number or a mailing label from a recent issue of *ALASKA GEOGRAPHIC®* to: Alaska Geographic Society, Box 93370, Anchorage, AK 99509. If your book is returned to us by the post office because it is for some reason undeliverable, we will contact you to ask if you wish to receive a replacement for a small fee to cover additional postage.

Contents

The Bering Sea: Northern Treasure House

By Richard P. Emanuel

Nestled between the continents of Asia and North America, bridging East and West, the Bering Sea is a realm of seeming contradiction. It is a domain of raging storms and stifling fog, of winter darkness and relentlessly advancing ice. It also teems with life, sustaining the richest marine ecosystem off the coast of the United States, and one of the most productive ecosystems on Earth.

At least 450 species of fish, mollusks and crustaceans inhabit the Bering Sea. Twenty-five species of marine mammals use the area, including a dozen or more kinds of whales. Eighty percent of the U.S. seabird population — 36 million birds of 35 species — live in or migrate to these rich waters.

The U.S. fishery here yields roughly half of the nation's fish production, worth more than $1 billion annually. Walleye pollock landings constitute the world's largest single-species fishery by weight. Vast populations of salmon spend their adult lives here and Bristol Bay's sockeye salmon fishery is the largest anywhere.

For thousands of years, these bountiful waters have sustained people as well. Even today, subsistence lifestyles closely bind Aleut, Yupik and Chukchi communities to the cycles and vicissitudes of the complex treasure house that is the Bering Sea.

Yet there are indications of a decline in this vast productivity, signs of an ecosystem in flux, or even distress. Some key species are experiencing dramatic drops in population, leaving scientists, fishermen, Natives and environmentalists alike unsure what lies ahead for the remarkable Bering Sea.

FACING PAGE: *Winter shore ice crowds Boxer Bay on St. Lawrence Island. Much of the ice that covers the region in winter forms in the Bering Sea itself rather than flowing down from the Chukchi Sea and Arctic Ocean. As the ice cover thickens, the Bering Sea coast loses its coastal characteristics and becomes more like an inland plain, enveloped in extreme low temperatures and swept by unobstructed winds. (Chlaus Lotscher)*

Geography

The Bering Sea encompasses 885,000 square miles, 1-1/2 times the area of Alaska. It is a relatively well-defined oceanic region, bounded on the west by the Russian Far East and on the east by Alaska. To the south, it is screened from the North Pacific by the great arc of the volcanic Aleutian Islands, including Russia's Komandorsky (Commander) Islands, a geologic extension of the Aleutian chain. In the north, a critical link to the Arctic Ocean is maintained through Bering Strait.

The Bering Sea is equally divided between abyssal deep and shallow continental shelf, with a transition zone of continental slope between the two.

When Russian explorers reached the far northeastern tip of mainland Asia in the 1600s, they found Chukchis. Here mother and child at right pose with a man wearing a suit of bark armor. (Illustration by Martin Sauer from Voyage dans Le Nord de la Russie Asiatique..., *courtesy of The Anchorage Museum of History and Art)*

The Aleutian Basin, in the southwest, is 12,000 to 13,000 feet deep. A broad continental shelf, richly productive, runs from Bristol Bay west and north through Bering Strait and into the Chukchi Sea at a depth generally less than 500 feet.

Bering Strait is 55 miles wide between Cape Prince of Wales, on Alaska's Seward Peninsula, and Cape Dezhnev, on the Chukchi Peninsula. Cape Dezhnev is the easternmost point in Asia, named for Semen Dezhnev, a Russian fur trader who explored the eastern tip of Siberia around 1648. Eighty years later Vitus Bering, a Danish mariner in the Russian navy, first sailed the sea that bears his name.

Bering had been commissioned by Czar Peter I, and later his widow, Czarina Catherine I, to explore the far eastern reaches of the Russian realm. In 1728, he traveled north from Kamchatka and discovered St. Lawrence Island. Sailing through what is now Bering Strait, inclement weather kept Bering from sighting the American continent. He ventured a short way into

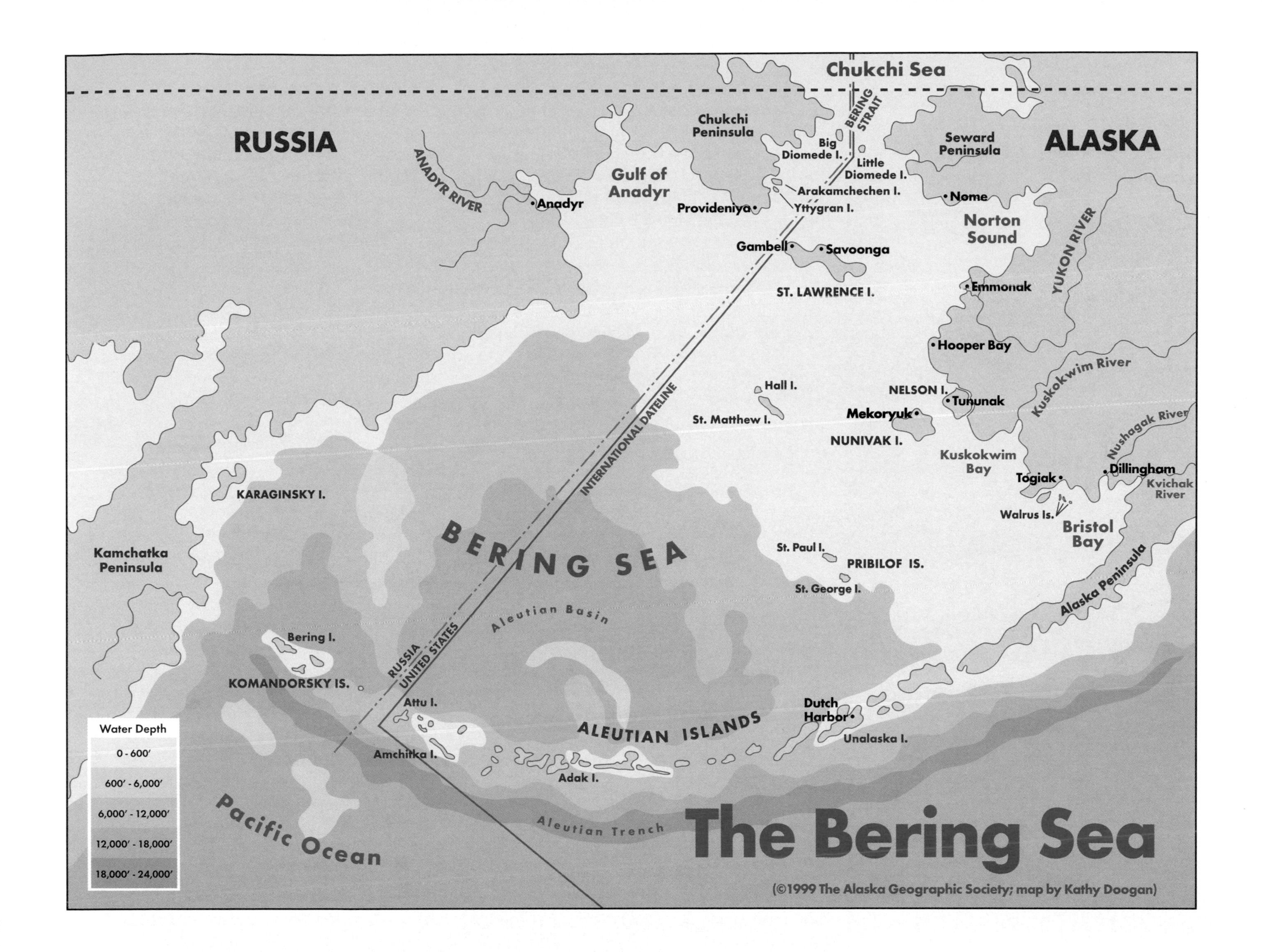

Chukchi Sea
RUSSIA
ALASKA
Chukchi Peninsula
BERING STRAIT
Big Diomede I.
Little Diomede I.
Seward Peninsula
ANADYR RIVER
Gulf of Anadyr
Anadyr
Arakamchechen I.
Yttygran I.
Provideniya
Nome
Norton Sound
YUKON RIVER
Gambell
Savoonga
ST. LAWRENCE I.
Emmonak
Hooper Bay
Kuskokwim River
Hall I.
NELSON I.
Tununak
INTERNATIONAL DATELINE
St. Matthew I.
Mekoryuk
NUNIVAK I.
Nushagak River
Kuskokwim Bay
Togiak
Dillingham
Kvichak River
KARAGINSKY I.
Walrus Is.
Bristol Bay
BERING SEA
Kamchatka Peninsula
St. Paul I.
PRIBILOF IS.
St. George I.
Alaska Peninsula
Aleutian Basin
Bering I.
RUSSIA
UNITED STATES
KOMANDORSKY IS.
Attu I.
Dutch Harbor
ALEUTIAN ISLANDS
Unalaska I.
Amchitka I.
Adak I.
Water Depth
0 - 600′
600′ - 6,000′
6,000′ - 12,000′
12,000′ - 18,000′
18,000′ - 24,000′
Pacific Ocean
Aleutian Trench
The Bering Sea
(©1999 The Alaska Geographic Society; map by Kathy Doogan)

the icy Chukchi Sea before returning south.

In 1778, the British dispatched Capt. James Cook to seek a northern passage between the Pacific and Atlantic oceans. Cook coasted the eastern shore of the Bering Sea, sailed through Bering Strait and onward to more than 70 degrees north latitude, about the area of Icy Cape on Alaska's northwest coast. He set course for Hawaii for the winter, but was killed there by Natives. His ship later returned to map the western shore of the Bering Sea in 1779, before proceeding to England. After this journey, the outline of the Bering Sea was broadly sketched by European explorers.

A young Little Diomede islander pulls a polar bear hide from its deep-freeze storage. The village of 178 Siberian Yupiks clings to rocky slopes on the west side of Little Diomede Island in Bering Strait. Its neighbor just three miles to the west is Russian-owned Big Diomede. (Al Grillo)

The Bering Sea has three major embayments: Norton Sound, south of the Seward Peninsula; Bristol Bay, between the Alaska Peninsula and the mainland; and the Gulf of Anadyr, south of the Chukchi Peninsula.

Three great rivers empty into the sea. The Anadyr River drains a sizable region in the Russian Far East and flows into the Gulf of Anadyr. In southwestern Alaska, the Yukon and Kuskokwim rivers have built a sprawling delta encroaching seaward.

A handful of important islands dot the Bering Sea, mostly igneous and volcanic in origin, although not geologically related to the Aleutian-Komandorsky volcanic arc.

On the Russian side, Karaginsky is the only sizable island north of the Komandorskys. It lies about 20 miles off Kamchatka Peninsula's northeast coast and is 60 miles long and up to 20 miles wide. Alder shrubs and a bushlike cedar, both typical of Kamchatka, grow on the island. These shrubs reach no more than 10 feet in height and are cut for firewood.

Koryaks, who lived in northern Kamchatka, used the island extensively until about the 1950s, when the Soviet government relocated them to the mainland village of Karaga to improve delivery of education and social services. Next, collective farmers tilled Karaginsky's soil, but the farms were abandoned toward the end of the Communist era. A 1993 visitor to the island found rusting farm machinery lying beside fallow fields, and Koryaks reoccupying summer camps, sailing small boats from the mainland and fishing from platforms anchored 300 to 500 feet offshore. But the only year-round residents of the island were the keepers of Karaginsky's lonely lighthouse.

The Pribilof Islands, in the southeastern Bering Sea, are home to millions of nesting seabirds. The islands also are breeding grounds for 80 percent of the world's northern fur seals; a smaller herd breeds in the Komandorskys.

St. Paul and St. George islands, each 10 to 12 miles across, were uninhabited

Scientists measure conductivity, temperature and depth with a CTD on the deck of the research vessel Alpha Helix. The CTD is the most important piece of oceanographic equipment on board, sending instant information about the salinity and temperature of the water to onboard computers. The rosette is made of several Niskin bottles, which can be triggered remotely to collect samples from any desired depth to study physical, chemical and biological properties of seawater. (Amy E.S. Schauer)

by humans until their discovery by Gavriil Pribylov, in 1786. Pribylov was drawn by sounds of bleating fur seals reaching him through the fog. Exploitation of the fur seals began immediately and the Russians brought Aleuts from the Aleutian Islands to the south to harvest the seals. The Aleuts gradually settled St. George and St. Paul and remain today.

St. Matthew Island, 230 miles northwest of the Pribilofs, also was uninhabited when found by the Russians, in 1766. The island is 35 miles long and 4 miles wide. Eight miles off St. Matthew's northwest tip lies smaller Hall Island. Both islands host seabird colonies but they are treeless, wind-whipped, ice-locked in winter and remain uninhabited.

A more substantial island is St. Lawrence, south of Bering Strait and 130 miles southwest of Nome. The island is 95 miles long, from 8 to 25 miles wide, and boasts two villages: Gambell, population 670, on the northwest tip, and Savoonga, population 632, on the middle north coast.

St. Lawrence lies closer to Russia than Alaska, just 40 miles off the Russian Far East coast. Islanders speak a Siberian Yupik dialect and share close cultural affinities with their neighbors to the west.

The sea north of St. Lawrence, where a branch of the Bering Slope Current bends north toward Bering Strait, is one of the most productive parts of the Bering Sea. It supports a rich benthic or seafloor community and a large population of Pacific walrus. For more than 2,000 years, St. Lawrence has been home to Eskimo hunting societies, sustained by walrus and the robust Bering food web.

Siberian Yupiks also inhabit the Diomedes, two small granite outcroppings in Bering Strait. The isles straddle the U.S.-Russian boundary. Big Diomede lies on the Russian side while Little Diomede, two miles away, is American.

The largest Bering Sea island is Nunivak, 60 miles long and 40 miles across, lying 20 miles off the Yukon-Kuskokwim delta. Nunivak was undiscovered by the Russians until 1821. Mekoryuk, population 192, on the north coast, is the only permanent settlement on Nunivak today. The island also hosts reindeer and a herd of muskoxen introduced in 1930.

A Lund skiff and a skin-covered umiak wait along the cobbled shores of St. Lawrence Island at Gambell, population 670. The aluminum skiff is usually used for hunting walrus; Natives take the umiak when they go after whales. (Jon R. Nickles)

Climate and Ice

For all the Bering Sea's riches, it does not offer humans a life of ease. One reason is the climate. In summer, the sea is ice-free and open water moderates temperatures. But dense fog forms when moist, warm air masses roll across cool open water. In fall, violent North Pacific storms veer northward over the Bering Sea with increasing frequency when winter storm tracks begin to take hold. As daylight fades and temperatures sink, ice forms on the shore, then steals across the sea.

This shorefast ice is a portent of things to come — still more ice. Some is floating pack ice from the Chukchi Sea, blown through Bering Strait by northerly winds. More ice forms in the Bering Sea in polynyas, open-water areas usually in the lee of islands and in the Gulf of Anadyr and Anadyr Strait. Bitterly cold temperatures freeze the surface of the sea in polynyas, but the ice is soon broken apart by wind and waves and the floes are driven out of the polynya, making room for more ice to form.

The process is "a wind-driven conveyor belt of ice," according to meteorologist Allen Macklin, of the National Oceanic and Atmospheric Administration. Wind keeps the polynyas open and feeds sea ice to the expanding body of floating pack ice. Where the growing mass of jostling pack ice meets shorefast ice, the once-open sea is transformed into a province of ice. The result is not a continuous ice sheet but a dynamic surface: Wind and sea fracture the ice into pieces and alternately pull them apart to form open leads or push them together to raise jagged ice ridges. The advancing ice is most fragmented near its leading edge.

For meteorological purposes, with the consolidation of sea ice, the Bering Sea coast is no longer a coast. It acts instead as a desolate continental plain across which polar winds may rage plunging temperatures, now unmoderated by open water, far below zero.

A herd of muskoxen roam Nunivak Island, largest in the Bering Sea and the last to be discovered by Russian explorers. (Gary Schultz)

Since 1990, populations of Steller sea lions, northern fur seals and some harbor seal groups in the Bering Sea have declined by between 50 and 90 percent. The result has been a crisis not only for the seals and sea lions but for the Bering Sea fishery, whose catch of pollock has been limited in an effort to promote recovery of the endangered sea lions.

Research by Don Schell, of the University of Alaska Fairbanks' Institute of Marine Science, doesn't reveal the reason for the marine mammals' decline. But it does provide critical clues and dramatically outlines the scope of the problem.

Schell, an oceanographer, has studied the ratios of carbon isotopes in whale baleen as a way to measure primary productivity, or growth of phytoplankton at the base of the food web.

Isotopes are varieties of a given chemical element that behave the same chemically but that vary in atomic weight, giving them subtly different physical properties. When a phytoplankton takes up carbon dioxide in the process of photosynthesis, some carbon isotopes are preferentially absorbed. The degree of the preference partly depends on how fast the plant cells are growing. The relative proportions or ratios of carbon isotopes in phytoplankton thus reflect how

Whale Baleen Documents Drop in Bering Sea Productivity

By Richard P. Emanuel

fast the plankton grew and the productivity of the waters in which they flourished.

Once incorporated into phytoplankton, carbon isotope ratios pass into the zooplankton that eat them. And when bowhead whales feed on zooplankton, carbon isotope ratios are preserved in the baleen they grow.

That's where Don Schell comes in.

Schell has examined baleen from more than two dozen bowhead whales taken in the Bering and Chukchi seas. Some baleen was from whales killed recently by Natives, other samples were located in the University of Alaska Museum in Fairbanks and the Los Angeles County Museum. The whales provided Schell with a record of Bering Sea productivity between 1947 and 1995.

According to Schell, carbon isotope ratios indicate "that seasonal primary productivity in the Bering Sea was at a higher rate over the period 1947-1966 and then underwent a general decline."

If his method is accurate, during the last three decades, the Bering Sea may have lost a third or more of its primary productivity. Recent estimates of the total mass of zooplankton in the region's waters support Schell's findings.

If the base of the food web is shrinking, there is less food for creatures at higher levels, and the implications for seals and sea lions near the top are clear. "This drastic decline ... is very likely implicated in the continuing decline of marine mammal populations," according to Schell.

Ever the careful scientist, Schell characterizes his findings as preliminary since colleagues have not yet thoroughly reviewed his work. The cause of the apparent drop in Bering Sea productivity remains mysterious in any case. Physical changes in climates or currents, subtle changes in salinity or seasonal ice or other factors could all contribute to new conditions in the Bering Sea ecosystem.

"Seeking the environmental, physical and chemical causes is the focus for future work," Schell observes. Meanwhile, his own carbon isotope studies of primary productivity continue. •

Fringed baleen plates protrude from a bowhead whale carcass at Gambell on St. Lawrence Island. (George Matz)

Ice-locked islands are likewise in this wintry grip, although in the lee of large islands, like St. Lawrence and St. Matthew, polynyas may remain all winter. Polynyas and other open leads are critical for the survival of marine mammals which need open water to reach the air they breathe, and resident seabirds which need leads to feed.

The timing and extent of sea ice coverage varies dramatically from year to year. As winter progresses, ice migrates southward. It typically reaches its maximum extent in late March or

Seaweed bends with the current crossing the shallow continental shelf near St. Paul Island. The current that sweeps the Bering Sea carries nutrients from the deep North Pacific, heightening the productivity of the mostly enclosed marine system. (Loren Taft)

Its reputation for bad weather firmly entrenched, the Bering Sea can be raked by fierce storms. These storms can be beneficial, however, when they stir the water column, making nutrients available to bottom-dwelling fish and crustaceans as well as mammals and seabirds feeding near the surface. (Dan Parrett)

April, when it may cover nearly all of the continental shelf. Sea ice usually stops short of the Pribilofs but stretches far down Alaska's west coast, reaching into and around Bristol Bay. Between April and June, the sea ice melts and retreats. The Bering Sea is normally ice-free from June until October.

The Puzzle of Productivity

Ocean ecosystems are intricate and hard to study, especially in an area as remote and with such inclement weather as the Bering Sea. As a result, the lavish productivity here presents an intriguing puzzle which scientists are only beginning to solve.

One piece of the puzzle has its origin half a world away, in the North Atlantic, where much of the water in deep circulation throughout the world's oceans begins its journey.

Cold water sinks in the North Atlantic off Greenland, explains Tom Weingartner, of the University of Alaska Fairbanks' Institute of Marine Science. It flows southward through the Atlantic, eastward around Antarctica and north across the Pacific Ocean. This takes hundreds of years and the water is receiving decaying matter from above as it moves along.

By the time North Pacific bottom water pours into the Aleutian Basin through passes in the Aleutian chain, it is as if the stuff has been spiked with Miracle-Gro™.

Circulation within the Bering Sea is counterclockwise around the Aleutian Basin. Along the continental slope there is a strong northwesterly flow, called the Bering Slope Current. Where it reaches the northern Russian coast, the majority of the current turns southwest into the basin; a smaller branch, driven by elevation difference, flows north, past St. Lawrence Island, through Bering Strait into the Chukchi Sea. On average, the surface of the Chukchi Sea is 20 inches below that of the Bering Sea.

As it flows up the eastern continental slope, the current draws nutrient-rich waters onto the entire eastern shelf, where the nutrients help maintain the complex cycles of life that characterize these shallow seas. In spring, the nutrients are taken up by blooms of phytoplankton, or microscopic plants, which are consumed by zooplankton, or microscopic animals. Zooplankton are eaten by larger organisms that are eaten by still larger ones, and so on up the food web. When an organism dies and decays, it releases its nutrients for recycling, a process most effective in shallow seas.

In the Bering Sea, wind and tides combine to mix the water column extensively, reintroducing nourishing ingredients into sunlit surface waters. Tidal effects agitate the bottom 130 feet or so of water. Strong winds are common in the region and may mix surface waters to a depth of from 30 to 100 feet, sometimes far deeper during a powerful storm. These actions stir the entire water column in areas such as the productive middle shelf — about half of the eastern Bering Sea — where depths are less than 230 feet. Such mixing promotes nutrient recycling. In the deeper outer shelf and along the continental slope, basin water loaded with phytoplankton is carried up by the Bering Slope Current, enhancing productivity.

Perhaps surprisingly, sea ice too spurs productivity. For one thing, many scientists have recently begun to appreciate how much grows in the ice itself.

"There are a lot of phytoplankton in sea ice in spring, called ice algae," Weingartner says. "Some people think that ice algae may substantially contribute to the productivity of the ice-covered Bering Sea and Arctic Ocean."

Another boost comes from the layer of cold, relatively fresh water that forms at the surface when sea ice melts in the spring. The water is fresher, or less salty, because when ice crystals form, salt is expelled. Fresher water is less dense, so it floats atop the water column, forming a thin, relatively well-defined layer at the surface of the sea.

Phytoplankton in this ice-melt layer are effectively confined within it, held near the surface where there is plenty of light for photosynthesis. The result is rapid growth, a phytoplankton "bloom."

"If the ice were not there, you would not get this bloom until much later," explains Vera Alexander, dean of the University of Alaska Fairbanks' Institute of Marine Science. "Such blooms occur in all seasonal ice-covered

Pink salmon collect in a spawning pool on Adak Island in the Aleutians. (Lon E. Lauber)

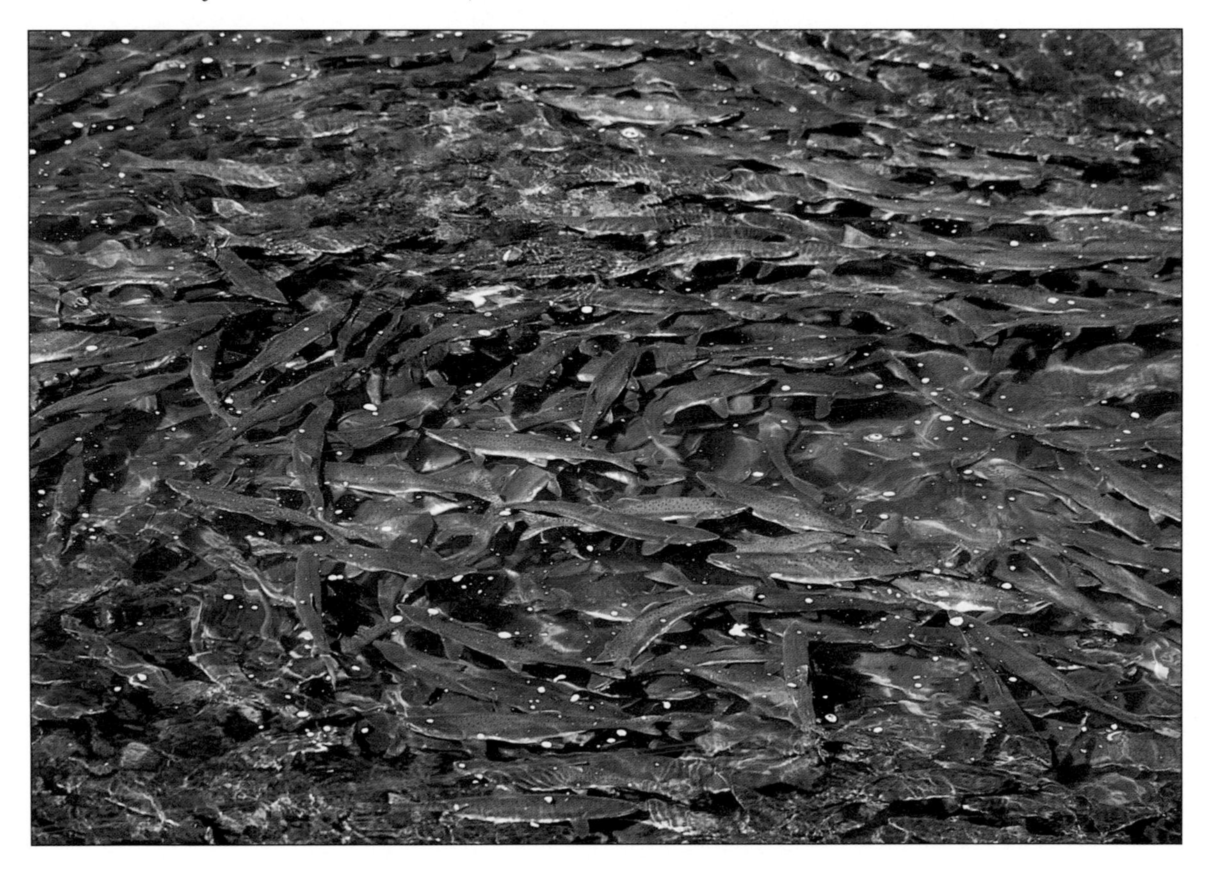

seas. We have compared the Barents Sea, the Greenland Sea, the Southern Sea," which rings Antarctica. "And they are all very productive. But nowhere have we found higher production than in the Bering Sea."

The early and vigorous bloom of phytoplankton rebounds up and down the food web. Zooplankton proliferate as the tiny animals feast on tiny plants. Larger organisms feed on the zooplankton and so on. As the bloom passes, the rain of decaying matter nourishes the benthic community: crabs and other crustaceans, clams. Eventually, walrus, sea lions, birds, whales, people all get into the act.

The elaborate workings of the Bering Sea ecosystem are at best only partially understood, but at least some of the elements for the northern sea's productivity do seem clear:

- the Aleutian Basin, filled with nutrient-rich bottom water, centuries in the making;
- the Bering Slope Current, drawn by outflow through Bering Strait, pulling basin water onto the shelf;
- an expansive shelf, shallow enough for good mixing by tides and by the region's strong winds and frequent storms;
- sea ice that jump-starts the ecosystem's powerful engine in the spring.

FACING PAGE: *A field research camp near Bull Seal Point offers temporary shelter for biologists studying treeless St. Matthew Island. The island lies 162 miles southwest of Nome and was first noted by Westerners when Russian seamen spotted it in August 1766, calling it St. Matthew. When he sailed the Bering Sea in 1778, British captain James Cook, not knowing that the Russians already had a name for the island, named it for Lt. John Gore of the royal navy, but the Russian name survived. (David Roseneau)*

RIGHT: *Yupik youngsters play on outboard motor packing material at Emmonak, population 838, on the lower Yukon River near its mouth. The Yukon flows 1,400 miles across Alaska from Yukon Territory before dumping into the Bering Sea. The Yukon combines with the 800-mile-long Kuskokwim River to create a tremendous delta 250 miles from north to south and at least 200 miles inland from the sea's eastern shore. (David Rhode)*

Change in the Bering Sea

Because natural systems are marked by change and dynamic balance, fluctuations are normal and may be temporary. But they are not always so.

In recent years, the North Pacific and Bering Sea have shown worrisome signs of change, though whether they are within normal limits or denote something is seriously wrong is hard to be sure.

Western stocks of Steller sea lions were listed as endangered in 1997 after an 80 percent drop in numbers over three decades. Harbor seals and northern fur seals have declined too, although fur seal numbers may again be stable. Fish-eating kittiwakes have fallen by 50 percent in the Pribilofs. Spectacled eiders are down 95 percent in their nest-ing grounds on the Yukon-Kuskokwim delta; eider numbers have fallen seriously though less precipitously on the North Slope and in Russia. In 1999, aerial surveys on the delta indicated the decline may be lessening.

Some scientists have implicated overfishing in the sea lion decline, and in 1994, a treaty closed fishing in the Donut Hole, an area in the Aleutian

Ice contributes to the Bering Sea's productivity by adding fresh water from melting and by providing a home for ice algae, a link in the food chain. This ice is breaking up in Norton Sound east of Nome. (Fred Hirschmann)

The port of Provideniya on Russia's Chukchi Peninsula anchors commerce in the northwestern Bering Sea. The piles of coal at left are used for heating buildings. (Amy E.S. Schauer)

Basin outside the 200-mile exclusive economic zone of both Russia and the United States. Waters around sea lion rookeries have also been closed to pollock fishing. Whether these steps will be effective no one yet knows.

Climate change is hard to detect because of large year-to-year variations in weather, but meteorologists have documented unusual conditions in the North Pacific and Bering Sea. They have defined a periodic shift in dominant weather patterns on a roughly 25-year scale, called the Pacific Decadal Oscillation, or PDO. In 1977, the Pacific shifted into a pattern of predominantly warmer weather, directing warm air toward Alaska.

In 1999, some scientists concluded that the PDO has shifted again and they forecast a decade or two of cooler weather for Alaska. The picture is complicated by long-range effects of El Nino and the possible onset of global warming, which is expected to show up first at high latitudes. Already, the arctic ice pack appears to be thinning.

Ramifications of these changes on Bering Sea life may be impossible to foretell. In 1997 and 1998, different sets of unusual weather and ocean conditions produced rare blooms of coccolithophores. Coccoliths are a kind of phytoplankton with calcium-carbonate shells. When coccoliths bloom, their shells reflect light, making it hard for some diving seabirds to see their prey. Short-tailed shearwaters died in vast numbers in the Bering Sea in 1997, and they were underweight in 1998. Coccoliths may be the reason.

Another example of ripple effects may be the 90 percent drop in western Aleutian sea otters since 1990. Hungry orcas, deprived of seals and sea lions, have begun to eat otters. Otters eat sea urchins, which have now been unleashed to devastate kelp beds, which support species from barnacles to eagles. Thus may the demise of sea otters wreak havoc with barnacles and kelp.

Despite these changes, the Bering Sea remains productive today. Recent poor salmon runs in Bristol Bay were not record lows. Escapements were met, allowing healthy spawning, and salmon stocks may well rebound.

LEFT: *Many of the Aleutian Islands are the tops of ancient volcanoes, whose bases were inundated when ice-age glaciers melted and sea levels rose. This rugged habitat allowed some remnant species of animals, such as the Aleutian Canada goose, and plants to survive in isolation. The Aleutian shield fern is found only on Adak Island, shown here. (Ed Bailey)*

FACING PAGE: *Shipwrecks provide an avenue for introduced species, such as rats, to reach Bering Sea islands, where they can wreak havoc with island ecosystems, killing native species and destroying habitat. This wreck is stranded on Unimak Island in the Aleutians. (Ed Bailey)*

Whale sightings in the Bering Sea have actually recently risen, perhaps an effect of PDO or El Nino warming. A northern right whale, once feared extinct, was recently seen in these waters.

For the first time, intensive, sustained scientific study is focusing on the Bering Sea. Meteorologists, oceanographers and marine biologists with the National Oceanic and Atmospheric Administration in Seattle have launched a multimillion-dollar effort to identify and track the components of systems that support life in the Bering Sea. Their work is coordinated with state wildlife biologists and academics in Alaska, Russia, Japan and elsewhere.

"Finally we are getting close to sufficient information to gain some understanding," says Vera Alexander.

To the geologically inclined, 13,000 years is a blink in time, but 13,000 years ago, Bering Strait was dry. Most of the continental shelf of the modern Bering Sea was exposed by a worldwide drop in sea level, as continental glaciers tied up vast amounts of water.

The resulting Bering Land Bridge connected Asia and North America and offered migration routes for land plants and animals and the humans who relied on them. Today, we rely on the shallow seas that inundate the shelf.

Humans may not halt global change but we can manage our own impact. Above all, we can hope to understand it.

It seems a wise beginning.

EDITOR'S NOTE: *Richard P. (Dick) Emanuel is an Anchorage free-lance writer who has contributed many feature articles to* ALASKA GEOGRAPHIC®. *He is also the author of* The Golden Gambell *and* Steve McCutcheon's Alaska, *both* ALASKA GEOGRAPHIC® *quarterlies.*

Marine Mammal Mysteries

By Kris Capps

The Steller sea lion population in the Bering Sea is declining drastically and no one knows why.

Northern fur seals, once numbering an estimated 2 million animals, have dropped by half and show no signs of recovering. Again, no one knows why. In recent years, Eskimo hunters at Gambell on St. Lawrence Island reported seeing fewer female walrus and calves. A scientific survey later confirmed their observation. Are calves not being born? Or are juvenile walrus dying? Nobody knows.

These are just some of the marine mammal mysteries surrounding the Bering Sea.

This incredibly productive ecosystem supports a variety of life: 450 species of fish, crustaceans and mollusks, 50 species of seabirds and 25 species of marine mammals, including whales, walrus, sea otters, seals and Steller sea lions. Despite recent protections for marine mammals by various international agreements, some species have undergone large and sometimes sudden population fluctuations. Steller sea lions have declined by as much as 80 percent and are listed as an endangered species. Under the Endangered Species Act, wildlife are labeled "threatened" when they are likely to become endangered in the foreseeable future. They are "endangered" when they are likely to become extinct in the foreseeable future.

Under the Marine Mammal Protection Act of 1972, a species becomes "depleted" when its population falls below optimum sustainable levels. This act prohibits the taking, i.e. harassing, injuring, killing or importing marine mammals, with limited exceptions for subsistence by Alaska Natives, scientific research, public display and incidental take in commercial fisheries; all require federal permits except subsistence.

Yet, marine mammal populations continue to decline. For example, since the 1970s, harbor seal populations have dropped by close to 90 percent in some areas. Scientists are trying to learn why these changes are happening.

Lloyd Lowry, Fairbanks biologist and chairman of the government's Steller Sea Lion Recovery Team, says he's not surprised answers remain elusive. "This is a complicated system and we haven't been studying it all that well for all that long. Even if we did study it for a long time, I'm not sure we could understand all the changes that go on."

It took alarm over the sea lion decline to focus widespread scientific attention on the Bering Sea. The

FACING PAGE: *A Steller sea lion bull and cows react to an intruder on Unalaska Island in the Aleutians. Alarm over the severe decline of western stocks of Alaska's Steller sea lion population focused the attention of the scientific community on changing conditions in the Bering Sea. (Dan Parrett)*

Sea otters, such as this one on Amchitka Island, have the densest fur of any mammal. Their coat is so thick that the underfur averages 300,000 hairs per square inch. As an animal ages, the hair on its head sometimes lightens. (Don Cornelius)

National Research Council, a private society of scholars in scientific and engineering research, formed a Bering Sea Ecosystem committee to study the region. In 1996 they issued a report pointing to environmental changes and human exploitation as the source of population declines of Bering Sea wildlife.

People have relied on Bering Sea resources since prehistory, and have exploited them commercially for the past 200 years by whaling, crabbing, sealing and fishing. In recent decades, population increases, especially among Alaska Natives on the Yukon-Kuskokwim delta, have put additional pressure on wildlife. Indigenous people on both coasts are not the only users of the sea. Commercial fishermen come from as far as Japan, Russia, Canada, Poland, Korea, Taiwan and China. Most scientists speculated the steep marine mammal declines were the result of overfishing of species eaten by marine mammals, climatic changes, pollution, disease, shooting by fishermen or entanglement in debris, including abandoned or discarded fishing nets. But when they examined all those possibilities, scientists determined that most did not play a major role in marine mammal declines.

Food, however, was the exception.

With the region's multimillion-dollar fishing industry in the spotlight, the Bering Sea suddenly became a political issue, the subject of lawsuits, congressional hearings, and the focus of an onslaught of scientific research. So began the drive to determine how things work in the Bering Sea and why things are changing.

Among theories scientists are exploring is one referred to as "regime shift." As it relates to marine mammals, the regime shift influences temperatures in the layers of waters inhabited by the mammals and their prey.

"This [shift] is something that happens on the order of tens of years," says Rosa Meehan, U.S. Fish and Wildlife Service biologist in Anchorage. "These are big climate patterns you could think of as long trends over time. There are other oscillations within that.

"At one regime you have overall cooler surface temperatures and less mixing," she explains. "Another regime, which is what we are into now, has warmer surface temperatures and greater stratification in water temperatures. That really affects the type of fish that are available near the surface. It also affects plankton distribution. That may be what is driving the abundance of various species out there." Although the regime may be shifting, the Bering Sea has been in that warmer temperature mode for quite a while now and scientists may be seeing results of that in marine mammal populations. While the theory is still speculative, Meehan says it is receiving growing support. "We are past the hand-wringing of the 'what's going on' stage," she said. "This is probably what is going on. It gives us

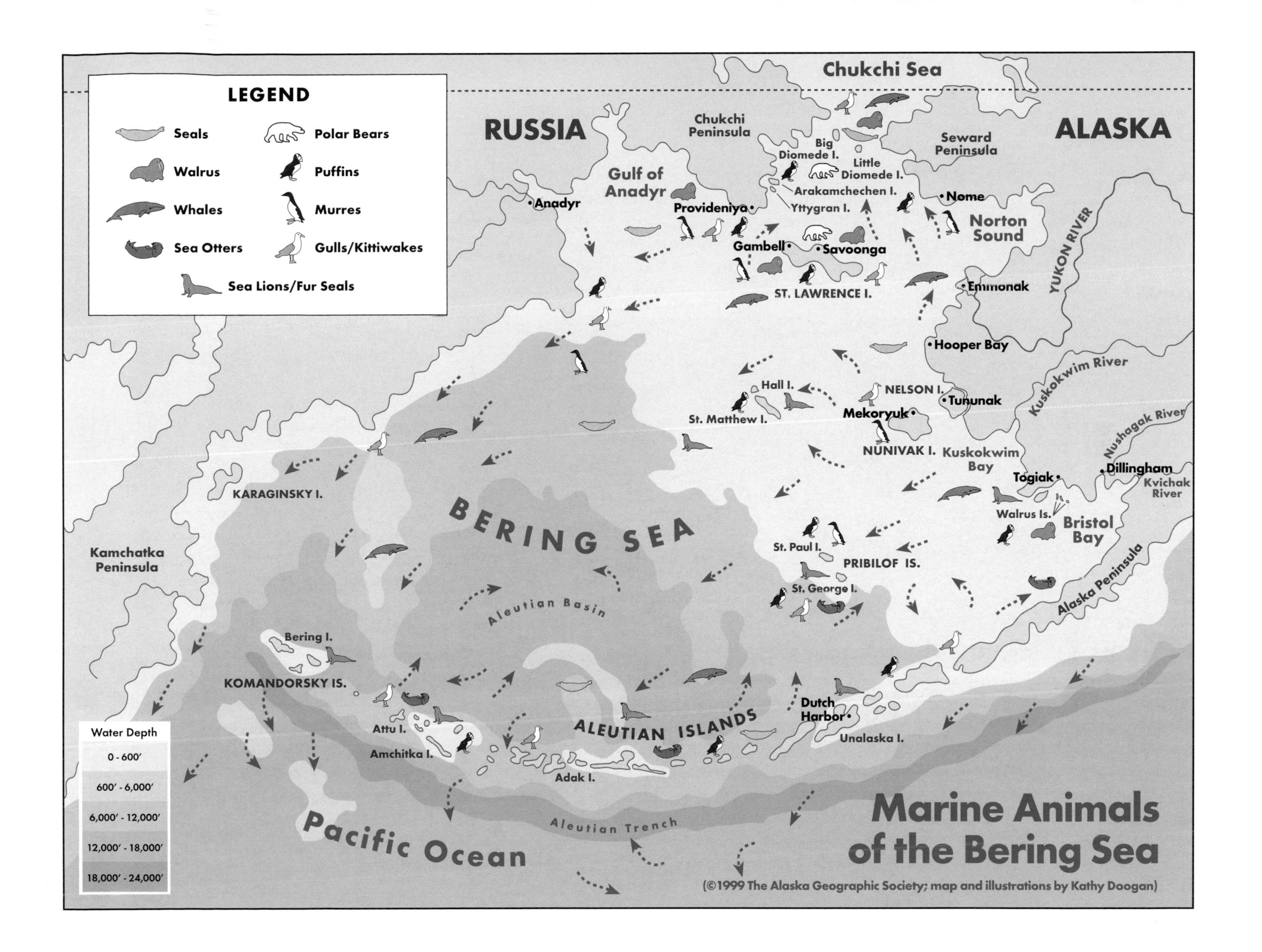
LEGEND
Seals
Polar Bears
Walrus
Puffins
Whales
Murres
Sea Otters
Gulls/Kittiwakes
Sea Lions/Fur Seals
Chukchi Sea
RUSSIA
ALASKA
Chukchi Peninsula
Seward Peninsula
Big Diomede I.
Little Diomede I.
Gulf of Anadyr
Arakamchechen I.
Yttygran I.
Anadyr
Provideniya
Nome
Norton Sound
Gambell
Savoonga
ST. LAWRENCE I.
Emmonak
YUKON RIVER
Hooper Bay
Kuskokwim River
Hall I.
NELSON I.
Tununak
St. Matthew I.
Mekoryuk
NUNIVAK I.
Kuskokwim Bay
Nushagak River
Togiak
Dillingham
Kvichak River
KARAGINSKY I.
BERING SEA
Walrus Is.
Bristol Bay
Kamchatka Peninsula
St. Paul I.
PRIBILOF IS.
St. George I.
Alaska Peninsula
Aleutian Basin
Bering I.
KOMANDORSKY IS.
Attu I.
Amchitka I.
ALEUTIAN ISLANDS
Adak I.
Dutch Harbor
Unalaska I.
Water Depth
0 - 600′
600′ - 6,000′
6,000′ - 12,000′
12,000′ - 18,000′
18,000′ - 24,000′
Pacific Ocean
Aleutian Trench
Marine Animals of the Bering Sea
(©1999 The Alaska Geographic Society; map and illustrations by Kathy Doogan)

a better framework to put things in." Other studies are beginning to link observations of birds, with marine mammals and with fisheries. "What we get out of this is predictability," Meehan explains. "We are able to put together this picture of what is going on."

Conducting research in the Bering Sea is no easy task, even though marine mammals might appear to represent a large target. In fact, they are inaccessible and often hard to find. Their range in the Bering Sea alone is 1-1/2 times the size of Alaska and sometimes overlaps international borders; weather in their home range is unpredictable and often nasty. While research focuses on certain animals, other species remain enigmas. For instance, there has been little research done on ice seals — ringed, spotted, bearded and ribbon seals — in about 20 years.

"What we know is not recent," said Kathy Frost, a biologist with the Alaska Department of Fish and Game's Fairbanks office, who studies seals throughout the state. "We hear a lot about Steller sea lion decline and seabird decline. We really wouldn't know if there is a problem or not with ice seals because there's just nothing to go on. There could be a decline of similar magnitude." Since research funding has not been allocated, Frost notes, "We haven't looked."

For a while, whales were in the research spotlight and most species seem to be doing well.

"Gray whales are at a historical high and bowheads appear healthy and happy," she said. "The whales here (in Alaska) that get attention are those that directly interact with people. Gray whales are real visible because they migrate close to shore and have so much visibility between Alaska and Baja." Scientists and coastal dwellers alike take notice when Bering Sea life begins demonstrating unusual behavior. That

The danger of being squashed by a bull is ever present for northern fur seal pups at the rookeries on St. Paul Island. Unlike true seals, fur seals and sea lions have external ears and can rotate their front flippers, making them far more agile over rocks and steep beaches. (Roy Corral)

Bull walruses are hauled out on Round Island. Both male and female walrus have tusks, actually elongated upper canine teeth. Females and young walrus usually follow the pack ice through the Bering and Chukchi seas. Bulls more commonly gather at seasonal haul-outs such as the Walrus Islands in Bristol Bay and in Russia's Gulf of Anadyr. (Fred Hirschmann)

has happened a couple times in the past few years, since scientists began paying more attention to this area.

In early 1999, hundreds of sea otters were reported starving to death on the Alaska Peninsula, blocked from their food source by an impenetrable wall of sea ice. An unusually cold February allowed sea ice to form farther south and a persistent high-pressure system kept it pinned against the shore. Some otters began a perilous overland migration, trying to reach the Pacific Ocean side of the peninsula through the Meshik River valley. The otters' plight soon became front-page news. Researchers, on the other hand, urged caution in placing too much overall importance on the event, saying it was actually not that unusual.

"This has happened a couple other times," said Lowry. "The distribution of ice determines the northern limit of sea otters in the Bering Sea. When we have a series of years of relatively light ice, they start creeping up. The ice pushes them back." Sometimes, he said, they

Robert Clark, of Clarks Point in Bristol Bay, and his family butcher a beluga whale that tangled in a subsistence net. The whales follow migrating fish into Nushagak and Kvichak bays. The light-gray bodies of numerous belugas surfacing in the muddy waters of the bay can be seen from the air when the salmon are running. (Greg Syverson)

get caught and killed by the ice. It is a "natural ecological phenomenon."

What is worth watching for, he says, is a long-term trend. For example, if global warming changed ice distribution, that would probably result in sea otters permanently altering their distribution as well. Currently the species is found in Alaska from Southeast to the western tip of the Aleutians and north into the Bering Sea to the Pribilof Islands, where they are spotted intermittently.

Not long ago, a research crew discovered another unusual event. In some areas of the Bering Sea, the sea otter population plummeted because otters were being eaten by killer whales. One theory to explain this behavior is: First, ocean perch and herring declined, either due to overfishing or sudden warming of the North Pacific, beginning in the late 1970s. This enabled a competing fish, pollock, to multiply. The pollock population grew at the same time whaling depleted the whale population. The main food source for baleen whales — microscopic animals like zooplankton — boomed and increasing numbers of pollock gobbled them up. Meanwhile, the more nutritious perch and herring declined. Pollock could not provide sufficient nourishment for Steller sea lions and harbor seals, so their numbers started dropping as well. Sea lions and seals are the major food for killer whales, and suddenly they just weren't there. Some scientists speculate that the whales then turned to the next best thing, sea otters.

Yet Lowry and others again urge caution in placing too much emphasis on these observations in terms of the Bering Sea's big picture. "Killer whales are intelligent pack-hunting predators," says Lowry. "They have the ability to do all kinds of different things. The fact they have this ability and have chosen to exercise it isn't very surprising. Whether it is connected to other shifts in overall prey availability for killer whales is speculation. It may also be just a pod that decided to have great fun.

"In 1990 or so, killer whales started

showing up around Naknek River eating beluga whales. It had never been documented in that area before. At the time, it was speculated that possibly this too had to do with the fact there were fewer sea lions available and they were ranging more widely to find other kinds of food. They did it for a couple years, then stopped.

"These are complicated interactions and we need to know the details of each one of them. To put them under the umbrella of things going wrong really does a disservice."

Lowry prefers to focus attention on analyzing truly unusual events, like the dramatic drop in numbers of sea lions. That, he believes, is worth worrying about.

The last aerial survey estimated 22,000 sea lions in western Alaska waters, down from more than 90,000 in the mid-1970s, a decline that spread east from Kodiak Island in the late '70s and early '80s, then west to the Aleutian Islands in the 1980s. Numbers continued dropping in the 1990s.

"I think it is unusual," he said. "It is too widespread, too big, too fast to be something you would expect a large mammal population to do."

Tom Loughlin, biologist with the National Marine Mammal Laboratory in Seattle, studies sea lions as part of the organization's Alaska Ecosystem Program. "The accepted leading hypothesis has to do with availability of food," he says. "And whether it is hard to come by because of changes in the environment — significant changes the past couple decades — or whether commercial trawl fisheries are taking the prey, or a combination of those things.

"Fishermen and some biologists they hire argue there is no observed or demonstrated link between fisheries and sea lions, so they shouldn't be managed in a way that hurts fisheries. We say even though the cause is unknown, it has been sufficiently shown that they compete for food. The pollock fishery takes their principal food. They feed in the same area, they get caught in nets, there is significant overlap.

"We think the conservative way to

King of a lonely domain, this Steller sea lion bull guards a stretch of cobbled beach in the Aleutians. Bulls average 9 feet in length but can grow to 13 feet and weigh 2,400 pounds; much smaller females reach about 7 feet and weigh 600 to 800 pounds. (John Hyde / Wild Things)

FACING PAGE: *A researcher observes Steller sea lions at a rookery during a study to monitor population trends. The western Alaska population of this species is officially listed as endangered. (John Hyde / Wild Things)*

RIGHT: *This sea otter looks forward to a meal of opilio crab in Unalaska Bay. This species eats primarily shellfish, octopus, and slow-moving bottomfish when their main prey is scarce. (Dan Parrett)*

approach this is to manage them in a way that causes the least amount of harm to sea lions," he says. Ensuing court cases will determine whether more research is needed to support that conclusion.

Overfishing is just one of the hazards marine mammals in the Bering Sea have endured.

Nearly all Bering Sea marine mammals have been hunted during the past 200 years, some heavily. The Steller sea cow was hunted to extinction in the 1700s and the northern right whale came close to extinction by the early 20th century. Mammals like Steller sea lions, harbor seals and species of ice seals have been consistently hunted for subsistence and for cultural reasons.

Here's a look at how some of the marine mammals are faring:

Steller Sea Lions

Before 1959, Steller sea lions were hunted, though only for subsistence, and scientists estimate Natives took about 400 animals a year. In 1992, a statewide monitoring program put the number at 548 animals, taken primarily in the Aleutian-Pribilof region. When commercial fishing boomed in the Bering Sea in the 1950s, humans began killing sea lions for other reasons. Research indicates about 140,000 sea lions were killed by all human activities between 1960 and 1990, including incidental takes and shootings to protect gear in the commercial trawl and salmon fisheries and wanton shootings associated with fisheries. Today in the Bering Sea and Aleutians there are an estimated 22,000 sea lions, down from more than 90,000 in the 1970s, with numbers steadily declining.

A harbor seal pup rests along the shore of the Alaska Peninsula. Harbor seal populations have held steady in the Bering Sea except for in a few spots such as the north shore of Bristol Bay and on Otter Island in the Probilofs, where numbers have declined substantially. (Ed Bailey)

Northern Fur Seals

From 1868-1911, more than 600,000 fur seals were taken from the Bering Sea. This hunting stopped with the Treaty for the Preservation and Protection of Fur Seals and Sea Otters, which expired in 1984. Since 1985, subsistence users have been the only fur seal hunters. Many fur seals were drowned in drift nets beginning in the 1970s, with more than 2,000 animals caught in some years. The population has dropped by half, to about 1 million. A subsistence season continues for fur seals on St. Paul and St. George islands. The most recent figures provided by Pribilof Islanders, from 1996, show a total of 1,823 seals taken, 1,591 from St. Paul and 232 from St. George. Only juvenile males are killed, to limit impact on the population.

Harbor Seals

From 1927 to 1967, state and federal governments paid a $3 bounty on harbor seals. The bounty continued a few years beyond 1967 in western Alaska. Although the bounty did not distinguish harbor seals from other kinds of seals, it is believed that annual kills statewide ranged from about 2,500 to 12,000 from 1927 to 1952. In the early 1960s, that number increased when the European fur market sold the skins. Estimated yearly catches statewide climbed from between 6,000 and 10,000 before 1963 to more than 50,000 in 1965. Many of these were pups taken from haulouts in Bristol Bay. Lower prices soon reduced catches to 8,000 to 10,000 per year. Subsistence numbers are unknown, but are estimated at 500 to 2,500 annually from 1970 to the early 1980s. In general, recent aerial surveys do not indicate a widespread decline in harbor seal numbers in the Bering Sea

Ice-associated Seals: Ringed, Spotted, Bearded, Ribbon

Annual catches for these species through 1972 were estimated at 15,000 to 21,000, with 65 percent being ringed seals, 20 percent spotted seals, 15 percent bearded seals and a few ribbon seals. Catch numbers declined after 1972 but current population figures are unknown.

Pacific Walrus

Walrus provide subsistence meat, oil for fuel and raw materials for other needs. Historically, skins covered driftwood for houses and are still used today to construct boats and ropes. Ivory was once carved to make tools, like harpoon tips. Today, the ivory is fashioned into handicrafts.

From 1931 to 1974, the walrus take in Alaska was thought to be less than 2,000 per year; 1974 through 1989 the take increased. The annual Soviet/Russian kill has varied from less than 1,000 walrus to more than 8,000. Walrus hunting remains an important part of the economy and culture of Native communities along the Bering Sea coast. The United States and

Females and pups crowd this northern fur seal rookery on St. George Island. About 80 percent of the world's population breeds in the Pribilofs; smaller rookeries occur in the Komandorsky Islands, in the Japanese Kuril Islands and at a few other spots. The subject of a lucrative fur trade off and on since the 1700s, which at times brought the species close to extinction, populations of fur seals in the Bering Sea seem to be stable in the late 1990s. (Dan Parrett)

Mature bull walrus reach 10 to 12 feet in length and may weigh 2 tons. (Harry M. Walker)

Russia have worked to monitor walrus, since their population ranges across international borders. The recent economic crisis in Russia, however, has diminished the monitoring program there, making Russian estimates no longer reliable.

Walrus were last counted in 1990 when their worldwide population was estimated to be about 200,000.

Sea Otters

Sea otters have been killed by indigenous people for thousands of years. They came under intense exploitation from about the mid-1700s, when Russian fur traders were active, until 1911, with an estimated 500,000 animals killed during that time. The Treaty for the Preservation and Protection of Fur Seals and Sea Otters ended that practice in 1911, when all hunting except by Alaska Natives was prohibited. From 1962 to 1970, the state of Alaska conducted an experimental harvest in which about 2,500 animals were killed on densely populated islands in the Aleutians. Since 1972, the only hunting has been by Alaska Natives for subsistence and production of handicrafts. In recent years the take has totaled several hundred animals per year. By the early 1990s, the estimated sea otter population for the Aleutian and Komandorsky islands and the north side of the Alaska Peninsula totaled 35,000.

Whales

Most whaling ended with international regulation in 1975. By that time, all large whales in the Bering Sea were severely depleted. Estimated total catches in the North Pacific from the 1950s to 1970s included 5,671 blue whales, more than 20,040 fin whales, 74,215 sei whales and 30,143 humpback whales — a significant portion of all baleen whales. Even those whales killed outside the Bering Sea, such as gray whales hunted in Baja California in the 1800s, might have affected the Bering Sea ecosystem because the whales are migratory. Gray whales have rebounded to an estimated population of 26,000. Bowhead numbers have grown to about 8,000 animals; beluga whale numbers are poorly known though estimates place 17,000 in Norton Sound and Yukon-Kuskokwim delta waters in summer. Northern right whales, once close to extinction, are still so rare their numbers hover in the hundreds.

As scientists study the Bering Sea, they are learning how the entire ecosystem works and the role each species plays. If Steller sea lions continue their dramatic drop, they could certainly become extinct, says Lloyd Lowry and others. They don't like to speculate on what effect that might have on the rest of the ecosystem.

"Certainly we would lose some diversity," says Lowry. "From an ecological standpoint, something else would fill the void left by the sea lion. There is probably no single ecological function of sea lions that would be missed, nothing we have readily identified."

Those who hunt sea lions for subsistence would certainly miss them, as

The hide of a freshly harvested bearded seal (oogruk) is tied to a snow machine sled on Nunivak Island. Largest of the western arctic and subarctic seals, bearded seals can weigh more than 750 pounds in winter and reach 7 to 8 feet in length. These animals depend on a thick layer of fat to keep warm in winter. (Gary Schultz)

LEFT: *A sea-going member of the weasel family, sea otters sometimes anchor themselves in kelp while feeding or resting. Adult males can grow to 5 feet and average 80 pounds. Smaller females average 4 feet and 60 pounds. (Dan Parrett)*

FACING PAGE: *Supreme predators, killer whales cruise the Bering Sea north of the Aleutians. Frequently called orcas for their scientific name,* Orcinus orca, *males average 23 feet in length, females are smaller. Bulls usually have a straight dorsal fin that can reach 6 feet in length, that of females and immature whales is shorter and more curved. (Daryl Binney)*

would tourists who spend thousands of dollars to see them in the wild. In past years, some species have disappeared due to evolutionary pressures and competition from other species and Lowry wonders if that is part of what is going on with the sea lions.

"It [the sea lion] may not be very well suited for the system the way the Bering Sea is set up right now," he says.

Perhaps all these marine mammal population declines are normal fluctuations, something that has been continuing, unobserved, for the past hundreds or thousands of years. Some scientists wonder if answers will ever be known.

It appears that the Bering Sea and the marine mammals that depend on it will continue to hold a fascination for scientists for some time to come. Tom Loughlin from the National Marine Fisheries Service describes the Bering Sea this way: "It's an enclosed area, it has high diversity, it is an intriguing biological area, and there are changes going on that no one has yet figured out. Intellectually, it is a very interesting place."

EDITOR'S NOTE: *Free-lance writer Kris Capps has written several articles on mammals and birds for* ALASKA GEOGRAPHIC®.

The Gray Whales Great Migration

By David Rugh and Alisa Schulman-Janiger

Gray whales have become a favorite target of America's whale-watchers because of their phenomenal migrations up and down the entire west coast of the United States and Canada. Much of this migration can be viewed from land or boats close to shore. Each winter these behemoths swim south thousands of miles from arctic and subarctic waters between Russia and Alaska to warm waters of Baja California, Mexico.

Not only are these migrations the longest of any mammal species, but they are also remarkably regular in timing. In December and January of most years since 1967, teams of observers working for the National Marine Fisheries Service have systematically recorded gray whales passing Granite Canyon, a counting station in central California. And every year since 1984, volunteers have helped the American Cetacean Society's Los Angeles Chapter count whales passing Point Vicente in Southern California from December to mid-May, covering both the southbound and northbound migrations. Records compiled from more than 20 years of full-season counts have shown that the peak (median) date of the migration varied by only four days.

However, there are puzzling differences in peak dates from one year to the next, and the peaks are now one week later than they were prior to 1980. A common explanation has been that changes in weather and the arrival of sea ice in far-north feeding areas is causing interannual differences in migratory timing. But arctic temperatures and ice-cover range too widely to explain migratory shifts limited to only a few days. A very cold day in late summer won't get the migration started, nor will a very mild fall prevent the migration from happening on schedule. We speculate that when whales inhale a blast of frigid Siberian air, or when sea ice prevails in the Chukchi Sea earlier than usual, the migration might be jump-started. But this may be only enough to escape the danger of ice covering the sea surface. When the migration is fully under way, it appears that whales slow or accelerate their speed to arrive in southern latitudes at about the same time each year.

An intriguing explanation for the regularity of the primary cue that starts migration is in the seasonal change in apparent day length. The shortening days of late October may trigger long-term memories, initiating the urge to migrate. Of course, the

FACING PAGE: *A gray whale breaches in the waters off Santa Catalina Island in Southern California. This view shows the animal's right side, with throat grooves visible along the upper right edge; the mouth; the eye seen at the base of the mouth line and white barnacle patches. (Alisa Schulman-Janiger)*

TOP LEFT: *In this close-up, orange whale lice cluster around a single gray-and-white barnacle. One species of barnacle and three species of whale lice colonize the skin of gray whales. Barnacles are scattered in patches, particularly along the top and sides of the whale. They are glued to the whale's skin, and do not move once the larvae have settled in from the water column. When the barnacles die and fall off, they leave a circular mark on the skin which, on healing, results in a white scar. The lice gather around the edges of the barnacles, the blowhole, genital area, anus and in throat grooves and skin folds. The legs of the lice have hooks that allow them to move around on the whale. (Alisa Schulman-Janiger)*

LOWER LEFT: *Unlike many great whale species, gray whales lack a dorsal fin. Instead, they have a low fleshy ridge down their back. (David Rugh/NMFS)*

length of a day will vary with date and with latitude. In late fall, whales at higher latitudes will perceive shorter day lengths than whales farther south on the same date. The farther north whales are in the fall, the earlier they might be cued to start south. Changes in the average location of the whole population in a given year may show up as changes in the average date that they pass researchers far to the south.

What then drives changes in the population's average location? Although most gray whales return to the same general areas each summer, particularly the northern Bering Sea and southern Chukchi Sea, where they feed within these areas may be inconsistent from year to year. Furthermore, a few whales disperse far to the west in the Siberian Sea north of Russia or to the east along northern Alaska, sometimes as far as the Canadian Beaufort Sea. Others spend the summer in the southern Bering Sea along the Alaska Peninsula. And some whales are seen along the southern coasts of Alaska or British Columbia or farther south in Washington, Oregon, California and even Mexico.

In pursuit of food, gray whales employ a unique feeding technique: Each whale turns onto its side on the sea floor and uses its huge tongue to create a vacuum in its mouth, sucking in sediment and prey, especially amphipods. Hundreds of short, yellowish, brushlike baleen plates hanging from the whale's upper jaws collect prey as the tongue compresses the mouth cavity shut. Each scour the whales make on the ocean floor can cover from about 2 to 24 square yards. Their impact on any one section of the ocean floor can be reduced if the whales move from place to place.

Ocean floor damage was not much of a problem through the earlier part of this century when there were very few whales. The excesses of commercial whaling almost drove this species to extinction in the mid-1800s and again in the early 1900s. However, the Eastern North Pacific stock of gray whales has made a remarkable recovery, becoming in 1994 one of the first stocks to be removed from the List of Endangered and Threatened Wildlife. In March 1999 a scientific review of the stock's status concluded that it was continuing to do well. Numbers have risen annually at a rate of approximately 2.5 percent, growing from a few thousand animals in the early 1900s to more than 26,000 currently.

The increase might have resulted in undocumented shifts in preferred feeding grounds. As whales seek new areas for forage, they may be at different latitudes when the days shorten and cue them to turn south. Perhaps too,

A gray whale shows its flukes as it begins to dive. Radio transmitters attached to individual whales indicate that the gray can dive to more than 500 feet and remain submerged for longer than 16 minutes. (David Rugh / NMFS)

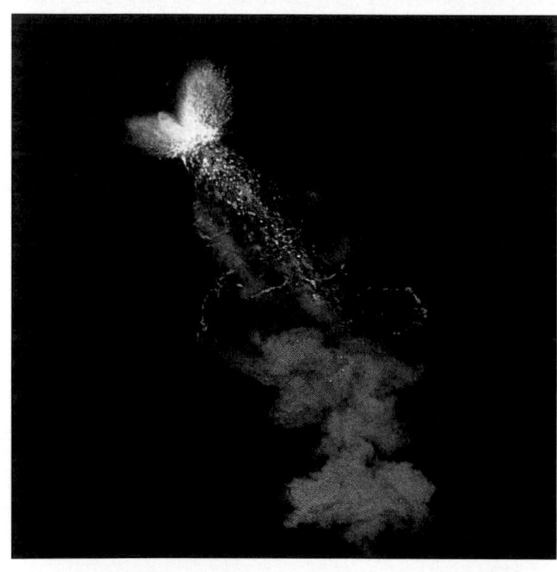

ABOVE: *A gray whale swims close to shore in this photo taken from the beach at Gambell on St. Lawrence Island. For decades, research indicated that gray whales bred and gave birth to young in lagoons off the Pacific coast of Mexico, but more recent observations point to other areas, such as the Bering Sea, as important for breeding. (George Matz)*

LEFT: *Sea bottom mud trails from a feeding gray whale. This species feeds by creating a vacuum with its mouth and sucking sediment and prey from the ocean floor. A gray whale eats about 200 pounds of food per day during the feeding season, about three to five months long. (Howard Braham / NMFS)*

whales are maximizing their time on the feeding grounds because food resources are not as plentiful now. This might explain the one-week delay in peaks of migrations after 1980.

An even more dramatic trend has been occurring in the arrivals of the migration. Since 1980 there has been an increasing delay, averaging one day every two years. Although the earliest migrants are scattered juveniles, the timing of the migration is set primarily by the large contingent of pregnant females. These whales make up the vanguard of the migration because they have the highest incentive to reach warm southern waters. Due to increased competition on the feeding grounds, however, whales might be taking longer to build up critical fat reserves needed to allow an animal to fast almost half of the year while swimming thousands of miles. In addition, females must store enough energy to deliver a 16-foot-long offspring and provide enough milk to keep it growing scores of pounds per day until it weans the following fall. If calving occurs at about the same time each year, which seems to be the case, then the later the migration, the farther north the calves will be born.

In the 1960s and 1970s few southbound calves were seen by observers in California but calf sightings picked up in the 1980s, first at the southernmost locations and later at the central California site. Now calf sightings are common. Last season's southbound calf sightings were especially high, consisting of more than 8 percent of the gray whales seen off Southern California. Calves have been seen as far north as Oregon and Washington. In fact, the Makah of Washington call the month of January *a-a- kwis-put'hl,* meaning

BELOW: *A 1-month-old, 14-foot-long calf swims beside its mother in Scammons Lagoon in Mexico. Gray whales can reach a maximum length of about 50 feet; the larger female averages 42 feet, the male 39 feet. (NMFS)*

RIGHT: *This close-up shows the head of one of a trio of whales trapped in an ice lead off Barrow in 1988. The wire leads to equipment for rescue and news crews drawn to the incident. Two of the whales reached open water after an international effort to save them. (NMFS)*

Sailing an umiak covered with walrus hide through drift ice, Siberian Yupiks from St. Lawrence Island seek bowhead whales during their spring migration. (Chlaus Lotscher)

the moon in which the whale has its young. This runs counter to the general perception of Mexico's lagoons being the calving and breeding grounds. Although some calves may be born there, it is now evident that many or most are born in January while they are still well to the north of Mexico. Also, breeding actually

Where To Watch Gray Whales

By David Rugh and Alisa Schulman-Janiger

The gray whale's migratory corridor is generally coastal, usually between the 100 fathom line and shore. Aircraft can provide good vertical views of migrating whales, but it is illegal to harass the whales, so aircraft must maintain minimum altitudes of 1,000 to 1,500 feet (or 2,000 feet over a National Marine Sanctuary) unless the pilot has a special permit. Boats can be ideal for closer inspections, but again boaters must not disturb the whales; 100 yards is the suggested minimum approach distance. There are concerns that intense whale-watching may be harmful to the whales. A more benign approach is to watch from shore, as thousands of people do each season. There are many good viewpoints along coastal highways all the way from Mexico to Washington state. Access to strategic coastal sites is more difficult in British Columbia and Alaska, except at Narrow Cape on Kodiak Island.

When should you expect to see the whales? The southbound migration takes about two months to go by; the northbound migration can take nearly twice as long. The average expected peak occurs on Dec. 11 at Unimak Pass, as whales leave the Bering Sea; on Dec. 18 at Kodiak Island; on Jan. 5 in northern Washington; on Jan. 8 in central Oregon; on Jan. 15 in central California; and on Jan. 18 in Southern California. The northbound migration is not as well documented, but records show there are two peaks: the first consists mostly of whales without calves, the second mostly mothers with newborn calves. These peaks pass the Los Angeles area on March 13 and April 22, respectively, beginning again the long migration to summering areas far to the north. •

occurs closer to the start of the southbound migration in the Bering Sea rather than in Baja. There are many reports of gray whales involved in amorous displays in the lagoons of Baja California, but these displays are also seen anywhere or anytime the whales are watched long enough. The effective syngamy, the moment that females become pregnant, occurs in late November or early December. Mexico's lagoons may play an important role as a warm-water nursery and resting area, but there is no scientific basis for describing them as the calving and breeding grounds.

In the winter of 1998-99 concerns arose about this stock of gray whales during its southbound migration. At some locations, such as in Oregon, almost no whales were seen when many had been expected in November and December. At other locations, such as on Kodiak Island, large numbers were seen in late December suggesting that the migration was significantly delayed. Based on these fragments of information, newspaper, television and radio reports indicated that this year's migration was very late; that it bypassed Washington and Oregon; or that it was stalled in Alaska. However, when the long-term records from systematic observations in California were compiled, it was determined that the southbound migration in 1998-99 was right on schedule. Once again, the regularity of the migration of these huge animals leaves us in awe.

EDITOR'S NOTE: *A gray whale specialist, David Rugh is a marine mammal scientist with the National Marine Fisheries Service in Seattle. Alisa Schulman-Janiger is the Gray Whale Census Project Director for the Los Angeles Chapter of the American Cetacean Society.*

A Growing Silence: Seabirds of the Bering Sea

By Penny Rennick

Their cries echo in the fog as I crawl through a forest of grass toward the high cliffs of St. George Island. When I find a place where I can peer over the edge, I see kittiwakes careening off vertical, volcanic ledges. Common and thick-billed murres adopt a more stately pose, standing upright in their tuxedolike garb. Horned puffins burrow into the tundra, their colorful bills signaling their identity. What in the Bering Sea's marine environment, I wonder, could support such prodigious numbers of wildlife? The bird cliffs of the Pribilofs and other vantage points in the region offer abundant vistas of seabirds, some humorous such as least auklets being forced backward by the wind as they launch from cliffs on St. Paul Island. Other images overpower, as when thousands of birds mass for flight before my eyes.

These visions of multitudes can be deceiving, however, for Bering Sea bird populations are declining just like those of many marine mammal species. Whatever is happening in the region is affecting the birds too.

Bird populations have not been changing uniformly. In many areas they've been decreasing, in other areas increasing. Red-legged kittiwakes, endemic to the region, have declined on the Pribilof Islands, but a smaller population on Bogoslof Island, 190 miles to the south, has increased. Since the Pribilof population is the main stock, however, the overall trend is downward. The same is true for many other species.

Bering Sea seabirds are facing the same situation as some of the marine mammals: How to feed their chicks and themselves when the pantry is undergoing drastic change.

Just what is happening to the seabird's food supply? Many current scientific studies and long-term data confirm that prey species that formerly made up the bulk of the Bering Sea biomass no longer do. Alaska Department of Fish and Game biologist Paul J. Anderson has kept records for more than 20 years on the catch from shrimp trawls in Pavlov Bay on the Alaska Peninsula. Even though these waters lie outside the Bering Sea geographically, they are linked because warm water from the Gulf of Alaska flows into the Bering Sea through Aleutian passes. Furthermore, both the gulf and Bering Sea are affected similarly by large-scale climate changes, and there is evidence for concurrent changes in marine communities. This data points to a shift from abundant stocks of forage fish —small, schooling species with flesh rich in oils and nutrients such as capelin —to pollock, a less suitable fish for the seabird diet because it contributes fewer calories.

FACING PAGE: *Tufted puffins, 14 to 15 inches long, renew their bonding after a skirmish. (Loren Taft)*

Yutaka Watanuki of the Japanese Institute for Polar Research and S. Dean Kildaw of the University of Alaska Fairbanks mount a radio transmitter on a common murre. Signals help researchers track where birds go at sea when they leave their colonies. (Jay Schauer)

Seabirds and sea lions target the oily, schooling fish, but there has not been a lot of study of these species because there is no commercial fishery for them. Anderson's data, undertaken to support the shrimp fishery, shows trawls from the 1970s hauled in capelin and herring with the shrimp; more recent trawls have brought up pollock but few shrimp or capelin. During a five-to-10-year period beginning sometime in the late 1970s, shrimp, once the source of a major fishery in Alaska, disappeared and they haven't come back yet.

In the Bering Sea, capelin similarly disappeared from diets of seabirds in the 1980s and the age of juvenile pollock consumed by seabirds changed also. This may have had a devastating effect on some seabirds because juvenile pollock do not provide the same energy value as capelin. They do not produce the fat levels seabirds, especially young, need. Consequently adults have to feed their chicks twice as much to produce the same amount of body mass and they have to work nearly twice as hard to catch the fish, which stresses them. Growth studies on puffins have shown that 100 grams of capelin can support the same body growth as 180 grams of pollock. Juvenile pollock are not bad for seabirds, just not as nutritious.

Another study in lower Cook Inlet in Southcentral Alaska, where capelin are still present, confirms the importance of oily fish in the seabird's diet. Near the Barren Islands, common murres feed their chicks almost exclusively capelin. Adults eat pollock. Adult murres may forage for hours in search of pollock, which they eat, then collect capelin that they carry home to feed their chicks. Pollock generates protein, capelin provides protein and substantially more fat. Adults try to load their chicks with fat to enhance their chances of survival in the critical first few weeks. If seabird chicks can make it through this period, they are looking at a 10- to 20-year life span. If chicks don't get enough fat, they become food stressed, a condition that can be measured by analyzing hormones in their blood. These studies have reinforced the conclusion that some percentage of oily fish is crucial to a seabird's diet. During winter and spring, this requirement can come from cod or pollock, which build up fat reserves in fall to sustain themselves through winter and perhaps as added insulation against cold.

Seabirds face a daunting task in collecting enough food with oil and fat to keep their chicks healthy. They must compete with adult pollock that cannibalize their young and with other large groundfish such as flounder and halibut that also consume large amounts of forage fish such as capelin

Several million short-tailed shearwaters congregated to feed north of Unimak Pass in the Aleutians in 1982. Since then shearwater numbers have declined, as have those of many other bird species that feed in the Bering Sea. (David Roseneau)

LEFT: *Another member of the alcid family that includes murres, murrelets, guillemots and auklets, horned puffins are named for the hornlike appendage above each eye that is shed after the breeding season. (Loren Taft)*

ABOVE: *Strictly a Bering Sea bird, emperor geese migrate from coastal breeding grounds to offshore islands, rarely flying beyond the region. Although the numbers of some seaducks have stabilized, populations of emperor geese are still down. (Dan Parrett)*

RIGHT: *In the same family as shearwaters and petrels, northern fulmars nest on the Pribilofs. The tube above the bill enables the bird to filter out salt from seawater. (Dan Parrett)*

ABOVE: *A pair of crested auklets and a pair of least auklets share a ledge in the Pribilofs. Cresteds are about 10 inches in length, leasts about 6 inches. (Patrick J. Endres)*

RIGHT: *Sunlight accentuates the colors in the plumage of this red-faced cormorant. This species lives year-round in the Pribilofs. (Patrick J. Endres)*

Common murres mass on Square Rock in Norton Sound. There are two murre species in Alaska, common and thick-billed. They can be distinguished by the delineation between the white underparts and the dark throat; in the common murre the line is rounded, in the thick-billed murre the white comes to a point. (Brad Stone)

and herring. And they must compete with commercial fishermen.

"It is not clear what the relationship is between commercial fisheries for pollock and declining seabird populations," says John Piatt, seabird biologist with the U.S. Geological Survey's Biological Resources Division in Anchorage. "Changes in seabird and marine mammal populations could result from natural variability in the food base."

Piatt thinks the change in seabird populations stems in large part from regime shift, the climate oscillation that last changed in the late 1970s and may be happening again. Climate data going back hundreds of years confirm this phenomena, a routine that is only briefly interrupted by El Nino and La Nina weather patterns. In the warmer regime since the 1970s, salmon, halibut and pollock flourished; seabirds, sea lions and crustaceans suffered. If the regime shifts back to colder conditions, seabirds and sea lions may do better.

Commercial fishing affects this balance, not by causing the shift but by accentuating its results, says Piatt. The Bering Sea pollock fishery is the world's largest. There may be just too many factors working against juvenile pollock, forcing the population down and keeping it there, and this for a fish that is not an ideal seabird diet to begin with. Piatt explains. "In natural systems, predators often have thresholds in their response to changes in food abundance. As long as food densities remain above threshold levels, and it is 'economically' profitable to feed on a particular prey, predators can fare well. Below those thresholds, predators either switch to other, more profitable types of prey, or starve if alternate prey are not available. In contrast, the profitability of fish for humans depends not just on fat or protein content, but on market conditions. Thus, the threshold for profitability in human fisheries changes constantly, and as fish become scarce they often increase tremendously in value, making it sometimes economical to continue fishing even for scarce populations. Unlike natural predators, then, humans are more likely to drive stocks to virtual extinction."

EDITOR'S NOTE: *Penny Rennick is editor of* ALASKA GEOGRAPHIC® *and a long-time bird-watcher.*

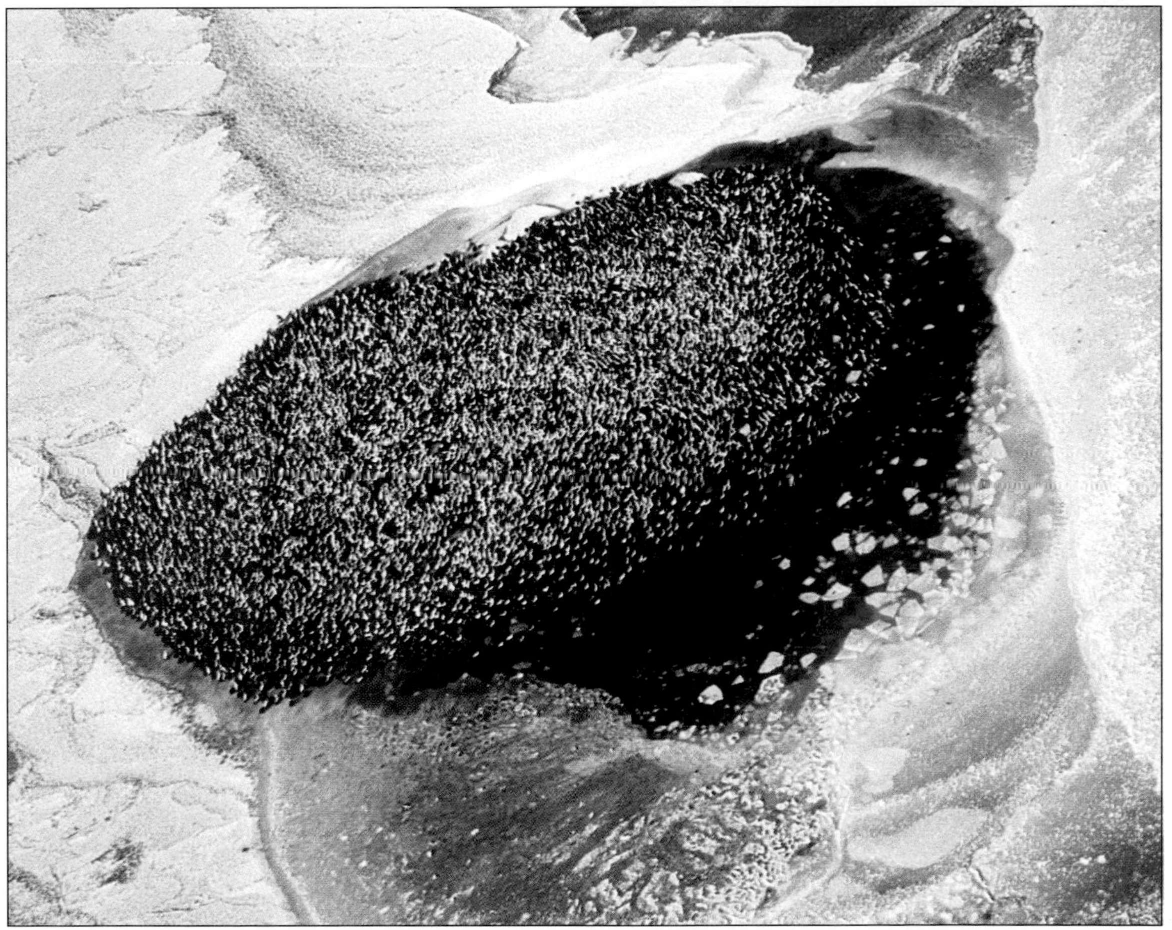

RIGHT: *Not until the 1990s did scientists know where some of western and northern Alaska's seaducks spent the winter. Through an unexpected resumption of a signal from a radio-tracking device, biologists were able to locate flocks of endangered spectacled eiders crowded into polynyas south of St. Lawrence Island. Subsequent research has found that some of these polynyas occurred over unusually rich benthic communities loaded with bivalves, such as clams, an important food for eiders. (USFWS)*

TOP RIGHT: *The decline of seabird populations in the Bering Sea and related waters may result in part from much lower numbers of a northern smelt called capelin, an important forage fish rich in oils that helps seabirds put on fat. (Greg Golet / USFWS)*

Troubled Waters: Bering Sea Fish and Fisheries

By Joel Gay

Looking down into the murky water along the eastern shore of the Bering Sea near the village of Togiak, it's hard to imagine the liquid holds any fish at all, much less the fabulous abundance for which the region is famous. Grayish-green and just a few degrees above freezing, the sea here looks less like water than cold, thin mud.

The nets tell a different story, however. In as little as 20 minutes of fishing, commercial boats will extract some 30 million pounds of Pacific herring from the turbid shallows around Togiak every May. The fish come flashing out of obscurity like electric lights, their silver scales the size and brightness of newly minted dimes. Net after net — some 600 boats typically fish each May — yields a trickle of herring that eventually combine into a stream, then a torrent, and finally a sea of fish bound for markets in Japan.

Just down the coast in June and July, gillnetters typically catch 100 million pounds or more of succulent, red-fleshed sockeye salmon in world-famous Bristol Bay. Working through winters dark and cold is the Bering Sea crab fleet, whose quarter-ton steel pots will lift out of the blackness more than 250 million pounds of gangly crab. Dwarfing even the 100-foot crab boats are the Bering Sea's trawlers, ships as long as a football field from a half dozen countries that fish nearly year-round for some 10 billion pounds of pollock, Pacific cod and other groundfish.

The amount of fish that comes out of the Bering Sea is staggering. Its cloudy waters give up half the fish caught in the United States, valued at more than $2 billion a year. Along its rim have sprung up villages, towns and cities, all dependent in some way on fish and fishing. Fishermen, soldiers, attorneys and engineers have battled over its riches, and hundreds of people have died in its cold, gray waves.

Though the Bering Sea may not be much to look at, fishermen from the United States, Russia, Japan and elsewhere have found that its assets more than make up for its lack of clarity. But the region is showing signs of stress. Not only are some marine mammals and seabirds in decline, some of its most valuable commercial fisheries have stuttered in recent years, causing many to wonder how much longer the

FACING PAGE: *King crab fishermen turned to other species of shellfish when the king fishery collapsed. This crew gathers in the hold of the F/V Amatuli with a haul of opilio Tanner crab. Tanner is the name used for the genus Chionoecetes, of which there are about five species in Alaska. The species commonly known as Tanner crab is* C. bairdi. *The name "Tanner" most likely came from Lt. Cmdr. Z.L. Tanner, commander of the Bureau of Fisheries steamer Albatross about the turn of the 20th century. (Dan Parrett)*

inscrutable Bering Sea can continue its breathtaking yield.

A diver opening his eyes beneath the surface of the Bering Sea would see little, his view clouded by the infinitude of plankton that sustain the entire food chain, and almost unimaginable numbers of salmon, pollock, crab, walrus and whales. The physical confines of the Bering Sea are responsible for its environmental wealth, says Pat Livingston, a National Marine Fisheries Service (NMFS) ecosystems biologist at the Alaska Fisheries Science Center in Seattle. The sea's broad, shallow continental shelf, the storms that rake it regularly and stir up essential nutrients, and the steady flood of summer sunlight combine to make it a thick, nutritious soup.

Unlike tropical seas with their profusion of brightly colored fishes, boreal ecosystems like the Bering Sea tend toward single-species domination. "These are simpler ecosystems," Livingston says, with fewer communities overall but higher production in each. In the Bering Sea, that dominant species is pollock.

Known both as walleye pollock and Alaska pollock, *Theragra chalcogramma* is thought to be the single most abundant fish in the world. It is found throughout the North Pacific, from Japan to California, but the bulk of the population lives in the eastern Bering Sea, the portion controlled by the United States. National Marine Fisheries Service surveys estimate the total weight of the pollock at more than 7 million metric tons, almost as much as the other major species of the region combined.

Dark gray above, white below and spotted along the sides, pollock typically weigh 2 to 3 pounds and grow to more than a foot in length. A small percentage of the soft, white-fleshed fish are filleted, but most are ground into surimi, a flavorless paste used as the base material for a host of seafood products, including artificial crab legs and shrimp.

Hardly any discussion of Bering Sea productivity doesn't touch on walleye pollock, a member of the cod family, that averages 13 inches but that can grow to 3 feet. Pollock spawn in masses, and tend to feed near the sea bottom, which is one reason they are a favorite target of groundfish trawlers. (Daryl Binney)

Among the most abundant fish in the Bering Sea, walleye pollock are caught by the tons. This species, especially juveniles, is also important prey for marine mammals, seabirds and other fish. (John Hyde / Wild Things)

Japanese fishermen, who first targeted pollock in their home waters in the 1950s, expanded into the Bering Sea in the 1960s. They were joined by large vessels from the Soviet Union, Korea and Poland, and by the 1970s the combined pollock harvest from the entire Bering Sea was estimated at around 5 million metric tons a year.

The foreign fleets' freedom ended in 1976, however, when Congress passed the Magnuson Fishery Conservation Management Act. The new law, enacted unilaterally, gave the United States control of all fisheries within 200 miles of its shore. Other coastal nations, including the Soviet Union, quickly followed suit, fencing off some of the most productive ocean in the world. But while fishermen now must observe boundaries, the fish they target do not, and migrating Bering Sea pollock have been especially hard to rein in.

The North Pacific Fishery Management Council, located in Anchorage, manages the federal fisheries off Alaska's coast, including three distinct pollock stocks. The main fishery harvests about 1 million metric tons a year primarily from the eastern Bering Sea, and NMFS biologists think it is stable. The other two stocks suffered marked declines in the 1980s. The Aleutian Basin area has been closed to fishing since 1991, while the Aleutian Islands area fishery has been pared back. The area was finally closed in 1999 after federal biologists said the action was necessary to protect Steller sea lions.

Russia controls out to 200 miles in the western Bering Sea. That region has its own pollock stock and fishery. Between these two domains, however, is a 41,000-square-mile triangle of international waters known as the "Donut Hole." Well after foreign vessels had been displaced from the American and Russian zones, trawlers from China, South Korea, Poland and Japan legally caught more than 1 million metric tons of pollock a year in the Donut Hole. Then pollock stocks began to decline in American waters. Scientists

Pollock are being emptied from the net into holding tanks on the Elizabeth F trawler during a summer pollock opening. (Daryl Binney)

using DNA samples soon confirmed that the Donut Hole catch comprised migrating American pollock. Using foreign aid payments as leverage, the United States and Russia crafted an international treaty between the six countries in 1994 that has since eliminated "dunking in the Donut."

Many in the U.S. fishing industry think boats fishing legally in Russian waters are again harvesting U.S. pollock. The fish, spawned in the southeastern Bering Sea, drift north as juveniles and commingle with Russian stocks. At about age 4 they swim back to the southeast, but they can live to be 15 years or older. Russia has embarked on a massive reconstruction of its aging fishing fleet, and is in dire need of the hard currency gained from selling fishing rights. At the same time, the country's faltering economy has forced cutbacks in scientific study and fisheries monitoring. U.S. fishery managers fear they may have to cut back on American fishing if Russia does not.

"It is clear that there needs to be a better basinwide understanding of stock dynamics and the effects of fishing," notes a 1997 report from the North Pacific council to Congress on the Russian Far East fishing industry. "A major challenge confronting the U.S. and Russia is how to work together to sustain this magnificent pollock resource."

As cod stocks in the North Atlantic decline from overfishing, world markets have focused their attention on another of the Bering Sea's treasures, Pacific cod (*Gadus macrocephalus*). Cod are larger cousins of pollock, typically weighing 8 to 15 pounds each, but the Bering Sea cod harvest is less than one-fifth that of pollock. Cod are caught by net, hook and pot, with much of the catch going to southern Europe as dried salt cod.

While pollock and other members of the cod family dominate the Bering Sea, the many species of flatfish (Pleuronectiformes) are a close second. Yellowfin sole (*Limanda aspera*) and rock sole (*Lepidopsetta bilineata*) are among the largest in both total number and individual size, and although they can be filleted, the fish are most valued for their roe.

Also targeted are sablefish or blackcod (*Anoplopoma fimbria*), Atka mackerel (*Hexagrammos stelleri*) and various rockfish. With the cods and flatfish, they are known collectively as groundfish. High in volume but low in value, most are caught exclusively by trawlers. Powerful ships up to 300 feet long, trawlers tow a net that looks something like a huge stocking. It is held open by a combination of floats, weights and steel plates designed to "fly" through the ocean and spread the net out. By varying the boat's speed, the net can be towed anywhere in the water column, including dragged on the bottom.

The boat captain finds schools of fish with sophisticated sonar electronics, then slowly tows the net through them.

When the tail of the trawl — known as the "cod end" — is full, the crew hauls it up a ramp in the vessel's stern using powerful winches and thick steel cables.

Trawl nets are wickedly efficient but indiscriminate, catching everything in their path. Fish that are the wrong size, species or sex are known as bycatch. With every cubic of inch of storage space a valuable commodity on a fishing boat, more than 600 million pounds of unwanted fish and crab is pitched overboard annually in the Bering Sea groundfish fisheries.

The Bering Sea fishing industry has developed so quickly that large factory trawlers can now catch groundfish and crab and process them on board. They come into port to offload their processed catch to foreign freezer ships that transport the catch to overseas markets. (Dan Parrett)

Facing public pressure to end the waste, Congress in 1995 instructed managers to reduce bycatch. Because of the size of the fishery, pollock was targeted first. In 1998 the fleet was ordered to keep virtually every pollock it caught, which reduced bycatch from 200 million pounds to 35 million pounds. Other so-called "dirty" fisheries will be targeted in the next several years to help reduce waste.

Many trawlers in the 100- to 150-foot range deliver their catch to shore plants in the Aleutian ports of Dutch Harbor, Akutan and other coastal communities. Even larger are factory trawlers, which not only catch the fish but process them on board. Individual fish are run through machines that cut off the head, remove the guts and fillet the carcass. Some boats have surimi equipment on board, while others freeze the fillets. Heads and guts are rendered into component parts — the oil is used to fire boilers on the ship, and the flesh is made into a dried, high-protein meal.

Unlike groundfish harvests off the coasts of Japan, Norway, Russia and even New England, the Bering Sea's catch has been relatively steady for 25 years. Biologists know fish populations rise and fall depending on environmental conditions, but federal managers have maintained conservative catch quotas, even in times of peak abundance. In addition, Congress enacted measures that cut the pollock fleet in half because of concern that the full fleet, with its ability to catch an excessive number of fish, and the ever-shortening season were putting unnecessary pressure on the groundfish fisheries.

That hasn't assuaged the fears of some that the Bering Sea fisheries may be in for a crash. They point to declines in seabird and marine mammal populations as evidence that the ecosystem is on the verge of collapse. Federal biologists and managers argue the case, saying factors other than fishing are causing those widespread problems.

Congress took control of the Bering Sea and other waters within 200 miles of the U.S. shoreline in 1976 out of concerns that foreign fleets were decimating American fish stocks.

Jasper Joseph pulls salmon from his gillnet on the Yukon River at Emmonak. (David Rhode)

The natural processes of the Bering Sea work in conjunction with feeder rivers to enhance the region's fresh- and salt-water productivity. Part of this cycle is the spawning and subsequent death of salmon that return to the region's rivers. When the salmon die, they become food for creatures farther down the food chain such as these sticklebacks. In the winter, rainbow trout, Dolly Varden and other charr feed on the sticklebacks. (Greg Syverson)

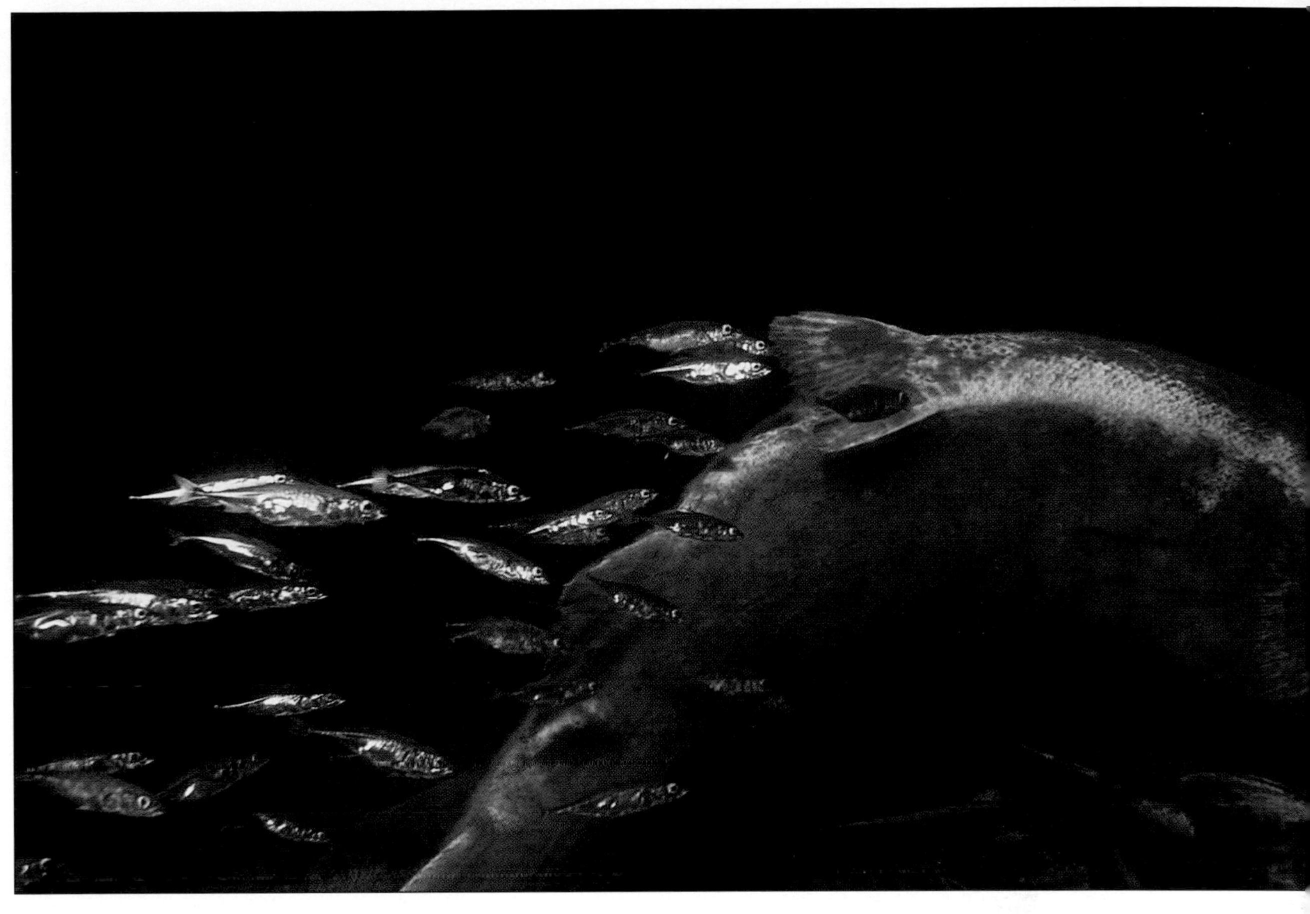

Ironically, the drive for statehood in Alaska was fueled largely by similar concerns, that salmon runs in the Territory of Alaska were being devastated by federal fishery managers in Seattle and Washington, D.C.

The first salmon processors arrived in Alaska in 1878. Within a decade there were salmon canneries throughout the territory, including on the Bering Sea coast. Under federal management the enormous runs of Bristol Bay and northwestern Alaska nearly collapsed, but they were nursed back to health and today are, for the most part, thriving.

Perhaps the most famous salmon run in the world is in Bristol Bay. The region hosts five of the six species of Pacific salmon but is best known for its sockeye. There are sockeye runs in Russia and throughout Alaska, British Columbia and the Pacific Northwest, but none to rival "the Bay." Its five major river systems together form the largest sockeye fishery in the world and are the backbone of the Alaska fishing industry.

Commonly called red salmon, *Oncorhynchus nerka* is a firm, red-fleshed fish that typically weighs about 6 pounds by the time it completes migration around the North Pacific. Adults spawn in river gravel; fry spend one to two summers in a lake on the same system, then swim to saltwater. Two or three years later they return as adults to spawn and start the process anew.

Just as the Bering Sea is blessed with certain geological traits that create favorable conditions for its high-seas resources, Bristol Bay seems to be made for sockeyes. Each of the river systems runs into a large lake at low elevation, and though the winters are cold, the long, sunny summers are balmy to hot — by Alaska standards, of course.

The murky Bering Sea saltwater that feeds the outmigrating sockeyes is eventually their downfall, however, as they can't see a net stretched out in

front of them. In the early days of Bristol Bay fishing, fish traps were used. After the traps were outlawed, fishermen were limited to using gillnets, a long strip of net with floats on the top and weights on the bottom that hangs like a curtain in the water.

A net deployed from a boat is called a drift gillnet; if it is anchored to the shore, it is a set gillnet or setnet. Both types of gear are allowed in Bristol Bay, though a fisherman must have a state-issued limited entry permit to fish there. The fishery was so lucrative in the 1980s that drift gillnet permits sold for as much as $250,000 a piece.

Prices have since declined to as little as $65,000, in part because sockeyes are not worth as much in the Japanese market, and because of a run failure. In the early 1990s, the annual harvest swelled to record levels, with fishermen catching more than 40 million sockeyes in each of several seasons. The long-term average had been around 20 million. Then in 1997 the sockeye return plummeted to less than half the expected number. The following year also came up inexplicably short.

Some fishermen think beluga whales may have eaten a significant number of returning salmon. Others think an increase in illegal salmon fishing on the high seas is to blame. Increasingly, however, scientists are convinced that environmental conditions in the entire North Pacific have changed, somehow reducing salmon survival. Warmer waters may have limited feeding

FACING PAGE: *The number one U.S. port by volume of catch and dollar value in 1997, the most recent figures available, is Dutch Harbor on Amaknak Island. Amaknak is connected by a bridge to Unalaska Island in the eastern Aleutians. The community of Unalaska, population 4,285, lies at center while various Dutch Harbor commercial developments line the left shore in this photo of Unalaska Bay. (Dan Parrett)*

RIGHT: *The Ocean Harvester catcher boat travels through ice during an opilio Tanner crab opening. Boat crews must monitor ice buildup on deck because severe icing can upset the stability of a boat and cause it to roll. Icing is a bigger hazard in the Bering Sea than in adjacent North Pacific waters because air masses over the North Pacific are warmed by the water, which is never colder than about 35 degrees, while in the Bering Sea northeast winds off the cold Alaska mainland can bring air temperatures well below zero, causing ice to build up from spray very quickly. (Daryl Binney)*

Crewmembers unload king salmon at a processor in the lower Yukon delta. The river fisheries of the Yukon and Kuskokwim have seen their take reduced as the overall productivity of the Bering Sea has declined. This effects not only the commercial fishery, but also makes it difficult for subsistence users to catch enough salmon to see them through the winter. Declines in both commercial and subsistence takes prompted the state to offer special payments to western Alaska residents who have endured two years of low salmon runs. (David Rhode)

grounds or food itself, increased predation may have killed juveniles — scientists are still sifting through clues.

Farther north, where the Kuskokwim and Yukon rivers flow into the Bering Sea, chum salmon runs also have sagged. *Oncorhynchus keta*, also known as dog salmon because of their use as sled dog food, are far less valuable than sockeyes because of their lower oil content and pale meat, and return in fewer numbers. Nevertheless, chums are both a crucial source of cash and valuable protein for the people who live along the rivers; when a run fails it is both an economic and social disaster.

The same can be said for king (chinook) salmon (*O. tshawytscha*). The big fish are high in oil, making them prized for drying and smoking. Failures in Yukon and Kuskokwim king and chum runs throughout the late 1990s led to the creation of a state task force to examine the causes and potential mitigation efforts of the missing fish. To date, there are many more questions than answers.

But while no piece of legislation can alter environmental conditions on the high seas or boost salmon survival, many Northwest Alaska fishermen think their lot could be improved by eliminating a fishery several hundred miles to the south. As salmon finish their meanderings through the North Pacific and head home to spawn, they swim through links in the Aleutian chain. The best known is False Pass, at the eastern end of the Aleutians, where for more than 100 years fishermen have gathered to capture sockeyes, chums and other salmon by the bushel.

Scientific studies show that a high percentage of those False Pass salmon are bound for Bristol Bay and the Arctic-Yukon-Kuskokwim drainages. The Alaska Board of Fisheries, which allocates the fish, has struggled to find a balance between False Pass fishermen and those farther north. Their decisions have been the focus of intense lobbying, courtroom challenges and even gubernatorial intervention.

The fishery at False Pass may get even closer scrutiny if the U.S. Department of Interior takes over management of subsistence fisheries. If that happens and Northwest Alaska continues to see low chum and king salmon runs, federal managers will have little choice but to shut down any fishery contributing to the decline. False Pass could be among the first targets.

The Russian Far East has its own runs of king, chum and sockeye, along with pink (*O. gorbuscha*) and silver (*O. kisutch*) salmon. In some respects, the fisheries operate much as they did in Alaska nearly 100 years ago, according to American observers in the early 1990s. Harvesting is focused mostly on terminal areas, at or near spawning areas. Fish traps are used widely still, which allows the salmon to be delivered to the processing plant still flapping.

The sockeye (red) salmon fleet gathers in Bristol Bay in what is still one of the most lucrative fisheries. Competition from foreign farm-raised fish and a glut on the world market have contributed to lower prices for Alaska salmon in recent years. (Joel Gay)

Siberians also use beach seine operations, in which a small boat deploys a fine-mesh net whose ends are brought to the riverbank, trapping the fish inside.

As Russia's economy opens up, increasing amounts of salmon have been sold on world markets, primarily in Tokyo, adding to the glut that has driven prices down for American fishermen in recent years.

Though pollock and salmon are the major industries of the Bering Sea, the infamous king crab fishery is probably best known. Alaska crab fishermen moved into the Bering Sea around 1970, after decades of learning how and where to find the wily crustaceans. Until then, they simply lacked the proper equipment to venture out where the crab were plentiful. Large steel boats made it possible.

When they first arrived, fishermen were searching for red king crab (*Paralithodes camtschatica*), and they found plenty, for a while. Catches peaked in 1980 at 130 million pounds, then collapsed. The fishery closed shortly afterward but since has rebuilt to sustain catches around 15 million pounds in a season that lasts less than a week.

With red king crab humbled, attention turned to smaller species. Catches of blue and brown king crab (*P. platypus* and *Lithodes aequispina*, respectively) amount to just a few million pounds a year, but these fisheries help keep the crab fleet busy by offering other areas and times to fish, from the Aleutian Islands to St. Matthew Island, beginning in August. Bairdi crab (*Chionocetes bairdi*), commonly known as Tanner crab, has been a lucrative fishery for the crab industry, though stocks declined in the early 1990s to the point

Sockeye (red) salmon fry hide in the vegetation of the Agulowak River near Dillingham before moving into the Bering Sea. After they mature at sea and return to spawn, these salmon will make up part of the world's largest sockeye fishery. (Greg Syverson)

Doug Herring hefts a red king crab on the F/V Bering Sea. The Japanese started commercial king crab fishing in the region in 1930. U.S. fishermen brought their trawl gear here in 1947. The king crab fishery in the eastern Bering peaked in 1980 at 130 million pounds. The next year it dived to 33 million pounds; by 1983 there was no commercial fishery. Although fishing has resumed on a limited basis and research indicates a slight population increase the last couple of years, king crab numbers remain depressed. (Dan Parrett)

that no fishing was allowed in 1999.

The real bread and butter fishery for crabbers these days is another Tanner crab, *C. opilio*, better known as snow crab. Weighing just a few pounds each, opilio are another high-volume fishery, with catches as high as 300 million pounds a year. Stocks are cyclical, however, and have fallen to less than 100 million pounds. Higher prices have generally accompanied lower production.

Though every fish population goes through periodic fluctuations in size, the crab stocks of the Bering Sea are among the most mercurial in the North Pacific fishing industry. Ecosystem studies show that king crab larvae are a popular food with pollock and even salmon, which suggests that during periods of high finfish abundance, shellfish stocks may be subject to distress.

That hypothesis seems to bear out off Alaska's coast. A 2.7-degree increase

in ocean temperature in the late 1970s spawned immense changes in finfish survival. It marked the start of a 20-year run of record salmon production in every area of Alaska, as well as high numbers of pollock, cod, halibut and herring. Concurrently, however, crab and shrimp populations began to die off.

Though no one doubts the natural cycles of fish populations, many wonder what can be done to even out the highs and lows, and whether fishermen and managers are somehow contributing to the roller-coaster ride. Answers may be forthcoming as Congress has approved additional federal funds aimed at unlocking some of the mysteries of the murky Bering Sea.

EDITOR'S NOTE: *Joel Gay, managing editor of the* Homer News *and a former commercial fisherman, is the author of* Commercial Fishing in Alaska*, Vol. 24, No. 3 of* ALASKA GEOGRAPHIC®.

ABOVE: *Bering Sea fish stocks support subsistence, sport and commercial fishing. Though this photo of herring drying in the community of Tununak, population 330, on the Yukon-Kuskokwim delta was taken three decades ago, herring are still an important subsistence resource, both as food and as bait. (Leonard Lee Rue III)*

FACING PAGE: *A catcher boat plies the Bering Sea during the opilio Tanner crab season. (Daryl Binney)*

Heart of a Halibut: Coming of Age in the Bering Sea

By Larry Merculieff

I knew the halibut on my hand line was large, probably a female, probably more than 5 feet long and weighing nearly 200 pounds. I could tell she was hooked by the lip and likely to come off if I was not alert to every movement. Carefully, I maintained steady pressure on the cotton line, using every part of my body to hoist the fish to the surface.

Halibut are among the strongest fish in the Bering Sea, known to fight so fiercely that inexperienced fishermen can be injured once a fish is aboard the small craft, between 14 and 22 feet, that we Aleut typically use around St. Paul Island, my home in the Pribilof Islands. Novice fishermen often let the halibut fight after it is landed before subduing it; but beginning as children we would always hear, "Respect the sea and the halibut, otherwise you can hurt yourself." Self-reliance, awareness and respect are only a few of the life lessons taught to us by the halibut and the Bering Sea.

I took my time as I hoisted the halibut from 150 feet off the rocky sea bottom, one-quarter mile offshore and 11 miles from the village of St. Paul, which is predominantly Aleut. The wisdom and lessons of my elders were guiding me now: *If she wants to fight, go with her energy; don't fight back. Honor her life force and she will know to give herself to you.* I knew that if the fish turned her head down, she would have more power through momentum than either I or the cotton line could manage, so I worked to maintain steady pressure.

Any hesitation by me now and the halibut would know instantly, causing her to swiftly turn down, perhaps ripping the hook from her mouth or breaking the line. Fishing with a hand line differs from a rod and reel; we tie a hook and sinker — weighing 2 or 3 pounds — to one end of a nylon or cotton line about 170 feet long then jig by moving the hand line up and down. I've seen all kinds of bait — aluminum foil, orange peels, bacon rind — but these days most people use herring or squid.

I could sense this was a powerful, wise, old halibut. She knew to conserve energy until the last death struggle — or an opportunity to escape. The fish did not fight on her way up, a way of acknowledging my skill. Suraj Ma Anand, who is not Aleut and was making her first halibut trip, gasped at the size of the fish. I knew I would need help to gaff so I instructed Suraj while I continued to slowly bring the halibut to the surface. *Do not let the nose of the halibut hit the air before you are ready to gaff,* inner voices of generations advised, *otherwise it will start fighting.* Because I had caught this fish adjacent

FACING PAGE: *Susie Angaiak of Tununak picks dried herring from woven lines of grass on Nelson Island. (Roy Corral)*

Larry Merculieff strolls his homeland on the island of St. Paul, where the only community, also named St. Paul, population 761, overlooks the southern shore. Merculieff has served in various leadership positions with the Pribilovians and in state government for more than two decades. (Roy Corral)

to a riptide area filled with large rocks, I had to watch the speed and direction of our drift while I labored to bring up the fish. Riptide zones can cause a boat to drift a half-mile in 10 minutes and the drift's direction can reverse in the same short time. To succeed with this fish, I had to shut out all distractions.

Finally the halibut was to the surface; I could see she was indeed female, 5-1/2 feet long, hooked only by the lip, just as I had felt when she first struck the line. One misdirected strike with the gaff now and she would rip the hook from her mouth and be gone. But before I could gaff, the boat rocked to the riptide's swells and the halibut's nose lifted into the air. The fish's muscular back arched immediately, her tail thrashed in the air, and she headed back to the bottom.

Keeping slight pressure on, I let the line go. I had to be completely one with this halibut if I were not to lose her. I had to know her intentions before she acted on them: Too much pressure on the line and she would be gone; too little pressure and she would be gone. I had to know exactly when the halibut was about to reach bottom so that she could be turned back up using her own momentum. Today people might call this "Zen" fishing — the ability to take action without interference of thought — but for me, it is the ages-old way of the Aleut.

I felt the halibut begin to turn. Gently, I increased pressure on my line, bringing her head back up, and continued hauling again. On the end of the line there was no struggle, only weight. The fish was conserving energy for when we would face each other again.

For Aleuts who have relied on the Bering Sea for more than 10,000 years, taking a halibut in the proper way is a ritual; mastery is a rite of passage into adolescence and ultimately manhood. For me, the Bering Sea is my history book and halibut fishing connects me directly with ancestors who, using lengths of strong kelp before the advent of cotton, would have sensed the energy of the halibut just as I did with my hand line. They loved the sea, just as I did. For my ancestors and me, the smell, taste and feel of this place in the middle of the Bering Sea are wondrous.

By the late 1700s, when Russian fur hunters first invaded our homeland, the Aleut's seafaring technology was the most sophisticated of any North American culture. Our people traveled tens of thousands of miles in high seas *iqyak* — kayaks — to Southern California, the Pacific Islands, the coast of Japan. I have built a traditional Aleut kayak, and I know its superiority stems from an ability to move with every nuance of the water — a design that shows profound understanding of the fast-changing Bering Sea, its 60-foot waves and lethal storms.

Like the kayak to the sea, I had to connect with the halibut before I could hope to bring the fish on board. Many experts talk today about "making the connection" when they mean everything from improved communication with a spouse or child to bettering one's exercise program. But when we Aleut talk about "connection" with a creature or the sea, we mean the mental, physical, spiritual and emotional bond with nature that is the foundation of traditional knowledge and wisdom.

Hoisting the halibut up, I thought back to how I got to this moment. I was 5 when I was first introduced to the seafaring ways of my ancestors. Clustered around a windburned fisherman known in St. Paul as "Old Man," we children watched as he cut halibut on the green grass next to his home. He had just returned with a load of fish caught from a 14-foot, New England-style double-ended dory powered by a 10 horsepower outboard. This halibut would feed

Using modern technology and equipment, Canar Sundown and others hunt seals on Scammon Bay, north of Cape Romanzof on the Yukon-Kuskokwim delta. (Roy Corral)

For decades indigenous people of the eastern and western Bering Sea were isolated from one another by political boundaries. But in the early decades and final decade of the 20th century, Natives were allowed to contact one another across the Ice Curtain. This woman was photographed before 1909 outside a dwelling in the Russian Far East. The stones keep the animal hides covering the dwelling in place during high winds. (Photo No. B72.27.135, The Anchorage Museum of History and Art)

village elders first, then his extended family and finally his immediate family, a traditional sharing based on respect.

That summer day, we were entranced as Old Man skillfully butchered the halibut for specific meals — bones for soup, steaks, cuts for fish pies. More fascinating to me was seeing what the halibut's stomach held: sandlance, octopus and small king, Tanner and horsehair crabs.

Suddenly Old Man cut out the halibut's heart, the size of an oversized walnut, and held it out to us, moving inside the circle of children. "Whoever will eat this halibut heart raw will always catch as much halibut as you will need whenever you go out fishing," he proclaimed. Startled, we all stepped back. Then, without thinking, I said, "I'll eat it!" And I did.

Without me knowing, Old Man initiated me into a tradition that determines who is ready to go to sea. Aleut wisdom taught that youth with courage enough to eat a raw halibut heart will make good students. The ritual reveals who is a risk-taker, willing to experience the unknown to learn new things. Our entire community reinforced these ways by offering learning opportunities to children and then rewarding curiosity. From the day I ate that halibut heart, my extended family and the village men would take me fishing whenever they had a chance.

A month after my encounter with Old Man, my first halibut gave itself to me on a hand line. I was on an outing with my father, John Merculieff, who lived his whole life on St. Paul. As is custom, I was required to consume the fish's heart, to become "one" with the halibut, and I sensed its spirit enter me the moment I swallowed. For hours as we continued fishing I gazed at the halibut that had chosen me, experiencing for the first time profound connection with a creature that had sustained my people for millennia.

Is it any wonder our ancestors viewed the Bering Sea as a divine being?

My first halibut went to extended family and one piece was kept for me to eat. Catching the fish was exhilarating for a 5-year-old but there was nothing like the delight and gratefulness in the faces of those who shared it: *Give away your first halibut and halibut will always come to you.*

The Bering Sea is a comprehensive school where one is taught not just how to make a living but how to live in harmony with oneself, family, community, animals and the earth. But as is tradition, I did not ask a lot of questions as a boy. I was encouraged instead to simply observe our fishermen and mimic them. In fact the number of words in this essay are more than those expended in all my years of learning to fish. Whenever there were words, they were filled with lessons: *If you do not have proper respect, you will die or you will not find food. Feel the texture of the water because it is different under different circumstances. She will give warnings because she is compassionate.*

I remember listening to my father and the other men speak in Aleut with a reverence for halibut and the sea. They would talk about the turning of the tide, or whether the bottom was "coming up" in fishing spots where the sea floor was basalt rock.

Awareness of the most subtle change in the sea bottom increases a halibut fisherman's success. Most halibut feed within 3 feet off the bottom although many times we have caught a fish 60 feet off the bottom if it followed bait on our hooks as we hauled up the lines. I learned that, depending on age, halibut forage in distinctly different bottom terrain; in fact, Aleuts know the sea the way city dwellers know the aisles of their local grocery store — 3-foot halibut in one fishing area, 4-footers in another, 5-foot fish some place else.

Children wearing marine-mammal-skin garments play among a selection of grass baskets in front of the store at Teller on the Seward Peninsula. (Photo No. B72.27.72, The Anchorage Museum of History and Art)

Through experience, and without benefit of compass or map, I learned the sea bottom topography within a six-mile radius of St. Paul, as well as

the three islands adjacent that make up our region of the Pribilofs.

I witnessed how men absorbed information through all the senses, how they could read the clouds, the color of the water, the direction and speed of drift. They knew the timing between tides, movement of wind, the sea bottom and the general movement of the sea. As a young fisherman amid teachers of few words, I began to understand the value of self-awareness and the need to stay connected to the sea, air and land to catch halibut — and to stay safe. I was learning to communicate in the ancient language that had allowed our people to thrive for hundreds of generations.

I went through my next rite of passage at 11 when my father gave permission to use his boat and motor

FACING PAGE: *Bertha Ohman picks wild celery on the Yukon-Kuskokwim delta. The shoreline vegetation is an important component of the Bering Sea ecosystem, holding the soil, providing nutrients and sheltering young animals before they go to sea. (Roy Corral)*

RIGHT: *Beach rye and other Bering Sea shore vegetation is woven into containers and platters for everyday housekeeping and as an outlet for displaying cultural artistry. Rachel Smart of Hooper Bay, here posing with her grandson, made this coiled grass basket. (Roy Corral)*

FAR LEFT: *Estelle Oozevaseuk, a storyteller and elder from Gambell, recalls changes that she has seen in her Bering Sea homeland. Traditional knowledge has enabled scientists to build a framework in which to assess current changes in the environment. (Al Grillo)*

LEFT: *Retired whaling caption Steven Aningayou describes the intricacies of traditional whaling to a 1991 visitor to Gambell on St. Lawrence Island. (Al Grillo)*

for halibut fishing. There was no competency test to earn this privilege; my father knew I was ready. By 11, I could navigate safely without a compass despite summer fog that coincides with the return of halibut from their southern migration. Seeking to avoid the cold, the fish move to deeper water just north or south of the Aleutian Islands, some 300 miles from St. Paul.

Even as a boy, I could feel, smell and read the texture of the sea and air and know when it was time to return to avoid a storm. I knew what part of the day halibut preferred to feed, where and when they go to give birth. I knew how to "ride" a skiff or dory when caught in large swells or breaking sea. I knew the sea bottom.

Although confident at a young age, I also retained respect for the Bering Sea and the halibut it held. The price of arrogance in the face of the Great Mystery — the divine mystery of creation — could mean death. I knew that even for the most accomplished mariner, there always was more to learn about halibut and the sea.

The way of the hand line allows a fisherman to feel the halibut directly, to communicate almost as if through a telephone line. Through the hand line, I can sense a halibut is near the hook even before it strikes and I can prepare for the lightning fast bite; otherwise, the fish will take the bait before I can set the hook.

Through the line's subtle movements, I can tell if the bait is being sucked in but not taken by the fish. I can tell if the halibut is merely moving its body across the bait before deciding whether to bite. Once on the line, I can tell how the halibut is hooked — by the lip, jaw or gullet, or snagged on the body. This knowledge guides how the fish is brought up — quickly or slowly, gently or vigorously. I can tell the size of the halibut and how much it will fight before ever hoisting my line. All this information helps determine success but it is unavailable to commercial fishermen who use longlines, in which 50 or more hooks are set on a line as much as three miles long and sunk to the ocean floor to be retrieved by a hydraulic winch — a technique that eliminates intuition or the use of one's senses. Young Aleuts today who use longlines without first learning the traditional ways lose much in their understanding of the fish.

Finally, after the fish surrenders, I bring the great halibut into our boat with Suraj's help. Her eyes well with tears. The halibut's end is filled with meaning; she has given her life with dignity, power and grace so that our lives may be sustained.

We spend the day butchering the fish, honoring it in the traditional way by making sure to waste no part. We return her skeleton back to the sea so that the fish will once again choose to feed someone else. We drum and offer a thanksgiving prayer. Even after some of the fish is given away to my extended family, there is still plenty to meet our needs for months.

After 44 years of fishing, the spirits of the halibut and Bering Sea continue to teach me, their Aleut apprentice. This wisdom is the true gift of the halibut heart, given to all who have courage to accept it.

Today the Bering Sea is in trouble: Populations of fish, seabirds and marine mammals — many taken for traditional food — are dwindling. On the decline are four species of eider; two species of murres; two species of kittiwakes; red-faced cormorants, northern fur seals; Steller sea lions; harbor seals; spotted seals; sea otters; herring; sandlance; capelin; and pollock, lucrative target of the bottomfish fleet.

If unchecked, these losses threaten the future of coastal communities — many of them Native — in a pattern I believe is not unlike the loss of South American cultures when rain forests are cut.

Native people, researchers and

A walrus skin is stretched to dry before being sewn into a boat frame for a Savoonga hunter on St. Lawrence Island. (Jon R. Nickles)

LEFT: *Using a traditional woman's knife called an ulu, a Yupik woman separates a fat layer from a seal hide. (David Rhode)*

FACING PAGE: *A Savoonga youngster holds a sculpin found in a tidepool along a St. Lawrence Island shore. Traditionally the fish could be used to supplement soup or as bait. (Jon R. Nickles)*

environmentalists are earnestly seeking ways to help the Bering Sea. There is much talk about the value of traditional knowledge and wisdom, which I believe can provide valuable insight into the Bering Sea's failing ecosystem and people's place in it. But scientists and others who genuinely seek to understand this way of knowing have a hard time because we believe it is connected to wisdom that comes from the heart, the bridge into the spiritual and the divine.

I am among a handful of Alaska Natives called to lecture and write about traditional knowledge and wisdom for educational or scientific forums and I am convinced there are many ways in which traditional understanding can complement science. For instance, it was Pribilof Aleuts who first noted that seabird chicks were falling off cliffs and dying in large numbers. The people also noticed that breastbones of some adult seabirds protruded and chest muscles appear caved in, both irregular conditions. Aleuts have reported that fur seal pelts are thinning and that more

Science Meets Traditional Understanding

By Rosanne Pagano

As scientists, conservationists and Alaska Natives ponder ecosystem shifts in the Bering Sea, more researchers are seeking to merge observations of traditional hunters and gatherers with those of biologists and other scientists. While formalized efforts to incorporate traditional knowledge are recent, informal efforts are not. Marine mammal biologist Kathy Frost, with the Alaska Department of Fish and Game in Fairbanks, says she encountered such information exchanges when she first arrived in Alaska 24 years ago. Today, give-and-take is an integral part of many research plans.

"In the Bering Sea, like elsewhere in Alaska, this sharing of information is important," Frost notes. "What Alaska Natives tend to know best is the area close to where they hunt, and that's where traditional knowledge can be incredibly valuable."

But traditional knowledge — like science itself — has its limitations. Covering 885,000 square miles, the stormy Bering Sea is so vast and remote that much of it is rarely visited by fishermen, hunters or biologists. To understand this big picture, new technology and long-term data are needed. Information collected and relayed by satellites and processed by heavy duty computers holds great promise, says Frost: "Trying to put this new information in context will depend not on high-tech advances, but on the shared insights of scientists, traditional people and others working together."

While striving for unadorned facts, scientists know too there are many ways of regarding an event — what Frost calls a "filter." For scientists, that filter is a search for the typical pattern, or norm, that occurs year to year and helps predict future events such as the migration of fish, birds or wildlife.

"But when you're living in a small village," Frost says, "everyone from age 7 to 107 knows what happens most of the time in their area. So the truly exceptional hunter, or the truly wise elder, is the one who remembers the odd years, the unusual way an animal sometimes behaves."

And while rare events or unusual years are complications for a biologist seeking the norm, "for a hunter trying to feed a family, remembering that once-in-a-lifetime event can become very valuable information," Frost notes.

A separate but parallel development has been the rise of "co-management" agreements, in which Alaska Natives who hunt marine mammals as traditional food share authority with government wildlife managers. In 1994, Congress formally endorsed co-management by amending the Marine Mammal Protection Act, authorizing the government to enter into cooperative agreements with Alaska Native groups to conserve marine mammals and co-manage subsistence harvests. Under the protection act, administered by the U.S. Fish and Wildlife Service and the National Marine Fisheries Service, only coastal Natives may hunt marine mammals. Co-management agreements today cover Alaska's sea otters, polar bears and walrus — all taken by Native hunters in the greater Bering Sea region.

Pursued by Natives desiring a greater say in the oversight of fish and wildlife, co-management formalizes cooperation among Native hunters, scientists and managers on issues such as safe harvests, scientific research and conservation. "The biggest difference now is that we're working with the hunters directly," says Rosa Meehan, an Anchorage biologist who heads the Fish and Wildlife Service's marine mammals management program. "Subsistence (hunting) is a very culturally based activity — it's more than simply gathering animals for the table. Co-management has provided a lot of insight both ways."

Frost, who is a charter member of the Alaska Beluga Whale Committee, says the mixing of traditional knowledge with science is especially gratifying when it prompts someone from either camp to reconsider findings and form a new, valid conclusion. For instance, some Norton Sound beluga hunters have learned enough about genetics research on population structure to contribute new questions and information, based on close observation of local whales.

"We can go a lot further working together," Frost says. "It's a synergistic sort of effect." •

EDITOR'S NOTE: *Rosanne Pagano is former associate editor of* ALASKA GEOGRAPHIC®.

seal pups are being preyed on by sea lions than anyone could recall.

These details told us that sea creatures were "food stressed," that is, something was amiss in the Bering Sea's usual predator-prey web; Natives observed this a full 10 years before Western science made an identical conclusion.

To the north, Inupiat Eskimo began noticing a thinning of the sea ice — an early indication, some experts say, of global climate change. Over the years other Alaska Native groups have contributed similar anomalous observations about the natural world — beluga whales spotted hundreds of miles from sea in the Yukon River, for instance. These reports are based on centuries of knowledge about the way things ought to be.

Diomede hunters work their way through rough ice on the Bering Sea. Russia's Big Diomede Island rises in the background. (Al Grillo)

ALEUT
CRUSADER of St Paul
ALEUT
CRUSADER
AK-1965-L

FACING PAGE: *In the past two decades or so, halibut fishing has developed into a viable commercial enterprise for Pribilof Islanders, but the traditional use of halibut has been embedded in the culture for generations. (Douglas Veltre)*

RIGHT: *Natives process a halibut catch at St. Paul. When the federal government phased out its participation in the killing of fur seals for their pelts in the early 1980s, the Pribilovians worked diligently to develop a fishing industry to offset the income and jobs lost from sealing. (Douglas Veltre)*

In the Bering Sea, Native people have thrived based on successful fishing and hunting that in turn depend on acute, informed observation. Reports from these Native groups also are useful because they are gathered year-round, not based on seasonal field work only. Native people can identify unusual behavior of animals at onset, as well as biological or population changes. Such information can help scientists form hypotheses that govern research — perhaps saving years of trial and error.

No single way of knowing is better but like biologists Kathy Frost and Rosa Meehan, I believe that traditional knowledge in union with science can be better than either world view alone. True, there is a vast divide between these ways of knowing that must be reconciled as we seek to heal the Bering Sea but as our elders say, nothing is created outside until it is first created inside.

EDITOR'S NOTE: *Larry Merculieff was born and raised on St. Paul and today divides his time between Chugiak, north of Anchorage, and his hometown. A past* ALASKA GEOGRAPHIC® *contributor, Merculieff is a coordinator with the Bering Sea Coalition focusing on the well-being of the sea and coastal cultures that depend on it. He served as Alaska's commissioner of commerce and economic development from 1988 to 1990 before returning to St. Paul for four years as city manager.*

People of the Foggy Sea

By Rosanne Pagano

In a canvas tent pitched on an Attu hillside, two men hunched over a hand-cranked recording machine set on a wooden crate. The man adjusting the cylinder phonograph's trumpet-shaped horn wears glasses and a suit; the other sits in a bulky jacket, hands on knees, intently addressing the machine which captures his words, spoken in turn-of-the-century Aleut. Under the tent's drawn-back flap sit three other men, perhaps waiting their turn at the horn.

Here is Waldemar Jochelson at work among the "people of the foggy sea," his term for the Aleut and title of a *Natural History* magazine article published by Jochelson in 1928. During a 19-month expedition in 1909-10, the bespectacled ethnographer who once plotted against a czar traversed the Aleutian Islands, hefting the phonograph recorder to the villages of Unalaska, Nikolski, Atka and Attu. The project, underwritten by a Russian banker, amassed volumes of archaeological, ethnographic and linguistic materials; 1,200 photographic negatives detailing Aleut life; and cylinder after cylinder of Aleut stories to share with anthropologist Franz Boas.

In 1943, some of the transcribed texts were left to the New York Public Library following Jochelson's death in 1937 and Boas's in 1942. The recording cylinders were archived in Leningrad, undisturbed despite revolution, world war and political upheaval. "I consider it nothing short of a miracle that this much has been preserved," writes Fairbanks linguistic professor Michael Krauss in his foreword to Jochelson's collection, *Aleut Tales and Narratives* (1990), published by the Alaska Native Language Center.

Jochelson arrived at a pivotal period for the Aleut, dominated by Russian colonizers in the 19th century and assailed by American culture in the 20th. "The Aleut people were struggling to uphold their traditions," writes Krauss, noting Jochelson's diligence not only in finding traditional storytellers but also in locating Aleut translators Aleksey Yachmenev and Leontiy Sivtsov, whose Unalaska home the ethnographer visited and photographed.

"This was a complex international enterprise that required money, planning, dedication, energy and rare vision," says Krauss, who has worked with the narratives. "For this Jochelson, no longer a young man, deserves our credit — as do Atkan Moses Dirks and the late Norwegian scholar Knut Bergsland, for transcribing the cylinders and editing the text."

Printed in Aleut and English, *Aleut Tales and Narratives* contains endless clues to a past in which the sea — its creatures and cruelty, bounty and beauty — is an intimate. Harvesting food from the sea is so familiar a task that the ocean's abundance is seldom remarked on directly, but may be gleaned instead from a narrator's sidelong observation. For instance, Nikolski storyteller Ivan Suvorov told Jochelson that when Aleut chief Qatxaykusax̂ wearied of "eating animal meat, he would have some of his men go fishing and then he would eat fish." For seagoing Aleut, the solution to gastronomic fatigue was simple.

In the lost world captured by Jochelson, the sea regularly offers up eiders, sea lions, seals

Ethnographer Waldemar Jochelson caught the Aleuts in transition when he visited villages in the Aleutian Islands in 1909-10. What he found were skilled seamen who routinely paddled stormy seas in kayaks to hunt sea otter, fur seal and other marine mammals. This pair is on a fur-buying trip. (Courtesy of The Anchorage Museum of History and Art)

Arrival of the Russians in the mid-18th century marked the beginning of drastic changes for the Aleut people. But not all Russians were set on subjugating their culture. The Rev. Ioann Veniaminov, who arrived in the Aleutians in 1824, diligently recorded Aleut ways even as he sought converts to Russian Orthodoxy. This photo shows Veniaminov's house in Unalaska. (Staff)

and whales. Even a husband can be found while beach combing. The tale "Usilax̂," recounted by Illarion Menshov of Unalaska, tells of a treacherous uncle who tries to kill his young nephew by hiding him in a skinned seal carcass bound at either end and abandoned at a seal rookery.

But the boy "began to feel that the tide was coming up and that he got afloat," Menshov says. "Thus he was conveyed along the sea until he felt that he had drifted ashore and heard two girls coming from his landward side, saying that they had found a seal." When the sisters undid the bindings, freeing the boy, "they said they had found someone to have for a husband (and) took him home...."

As any fisherman knows, the sea is not all goodness: Isidor Solovyov of Akutan cautions in the short narrative "What is not given to eat to a boy who is growing up" that fish tails bring on trembles while consuming the bottom part of a codfish bladder causes weak fingertips. Youth also should refrain from rock cod, known as Irish lord, so they "may not become old early."

In the account, "How the Aleuts hunt and fish," Arseniy Kryukov of Umnak offers a catalog of practical advice: "If a man who prepares to hunt sea otter goes first down to the beach at low tide and rubs himself with (kelp or seafood), the sea otter does not shun him. It also does not shun the one who gets up early in the morning and walks along the seashore....

"(We) catch fish: halibut, cod, sculpins, pogies, great sculpins, flounder, black bass," Kryukov adds. Pogie fishermen worked with a gaff among the rocks, while oceangoing fish were taken on a line. To land halibut, Aleut used a device Kryukov calls a *yarus,* a float made of inflated bladder tied to a line and fixed with bait. "Thus the halibut that has taken the yarus in its mouth is choked and dies."

Finally, no people who live and die by the ocean's moods can overlook its mysteries. In the three-paragraph tale, "Things that occur at sea," told to Jochelson by Solovyov, paddlers are advised that a reef appearing suddenly at low tide can be made to vanish if it is hit by a thrown spear.

"If at sea something like a swaddled child appears, we paddle to it and, passing by it, pull it out from its wrapping and (so) get safely away from it," Solovyov says. "That is the one called Aĝangix̂. However, whoever flees from it will get taken by it." •

The Spirits of Whale Alley

By Downs Matthews

Arakamchechen Island may not loom large in your personal geography. Few Americans, and for that matter, few Russians, know of this wildlife haven rising out of the Bering Sea beside the Russian Far East's northeast coast. Except for the summertime presence of two Russian wildlife officers, it is uninhabited. But there was a time when Arakamchechen and other spots nearby were home to hundreds of aboriginal hunters known to anthropologists as the Whale Alley people.

To find Arakamchechen, first locate the port of Provideniya, at the southernmost tip of the Chukchi Peninsula. From there look about 75 miles to the northeast to Arakamchechen Island. You'll find it separated from the Russian mainland by the Straits of Seniavin, otherwise known as Whale Alley, where bowhead whales once thronged.

Roughly a triangle, Arakamchechen has a surface area of about 60 square miles. Its hills and low mountains end precipitously at water's edge. Beaches are few but bird cliffs are plentiful. According to local tradition, the Chukchi hold sway over Arakamchechen. Yupik speakers from nearby Cape Chaplino worry a little about going ashore there.

Just opposite Arakamchechen to the south lies a smaller island called Yttygran, also an important Whale Alley community in ancient times. A few miles to the west, a jagged lump of rock rises out of the Bering Sea. Yupik speakers call it Nunianeghaq, or Small Island. It is seabird heaven. Between them, they define Whale Alley and its feeding grounds for bowhead and gray whales. Whales pass through here on their spring and fall migrations to feed on amphipods living in the bottom mud of the straits.

When Russian geophysicists mapping the Chukchi Peninsula in the 1940s and 1950s reached the Bering Sea coast, they came across dugout dwellings roofed with whalebone and other artifacts. They reported a score of these ancient settlements from the Gulf of Anadyr north past Cape Dezhnev, a distance of about 200 miles.

Russian anthropologists, alerted by the geophysicists, came to the area in 1976 and found an extensive ritual complex created from the bones of bowhead and gray whales. They concluded that around the first century, hunters of the Old Bering Sea culture perfected skills and technologies necessary to take bowhead and gray

FACING PAGE: *A spit and low hills delineate Yegor Purin Bay on the Straits of Seniavin side of Arakamchechen Island. On the side of the hill overlooking the cove is a concrete cross to which is affixed a copper plate with the following inscription: Yegor Purin, Secretary of Clipper Ship Guy'd'amak, 30 of August 1875. (Downs Matthews)*

Whalebones frame this image of Arakamchechen Island Nature Reserve Inspector Anatoly Sergeivich Gayevskiy (in middle) showing visitors a whale vertebra. Ritual sites marked by marine mammal bones are spread along 200 miles of the Russian coast on the western side of the Bering Sea. (Downs Matthews)

whales as food. Settling near the Bering Sea meat market in a score of widely spaced communities, they remained active for the next 1,700 years. Their numbers can't be known, but on Yttygran alone, experts estimate that as many as 170 Whale Alley people lived.

Using toggle-headed harpoons carved of ivory, braided ropes of walrus leather, floats made of walrus stomachs and walrus hide boats, hunters would attack and kill calves and smaller adult whales as they migrated through Whale Alley. Men and women would beach and flense the animal, saving the meat in lockers dug below the permafrost. Petroglyphs depicting whaling scenes show how they did it. In fact, some of these meat lockers still contain frozen whale meat placed there 300 years ago.

Anthropologists have concluded that Whale Alley served as the region's center for the ritual worship of whales. Elsewhere, bones were either discarded or used as structural elements in dugout homes. But on Yttygran, residents used rib, mandible and head bones to build a huge sanctuary for whale worship. The skulls of whales were buried in pairs, nasal parts down, with the occipital parts raised. More than 60 skulls form an arcade extending along the gravel strand for a thousand feet. Long mandible bones were raised like flagpoles at many spots. A shrine within a stone amphitheater was the stage for rituals and feasts.

Towering mandible bones raised on prominent points may have served as signposts for the paddlers of umiaks making their way home in fog and snow. Another theory suggests that bones lined the shore so that the spirits of those whales could call living whales in from the sea. To this day, Cape Chaplino Eskimos believe that the remains of any animal that lived in the sea must not be returned to the waters because their spirits would warn away living animals. Some 1,500 whale skulls have been counted from north to south along 200 miles of Chukchi Peninsula shoreline.

Whale Alley people buried their dead on nearby hilltops with graves positioned so the occupants could look out over their watery hunting grounds. They marked graves of men with skulls of sea lions or walruses or polar bears to demonstrate their prowess as providers. Graves of women were marked with symbols of their domestic skills, such as cook pots, needles and carving knives.

Among the Cape Chaplino Eskimos, the Whale Alley people are called the Masigmit, and they deny any relationship with them. Their folklore contains no explanation for where the Masigmit came from or where they went. But from the time that Russians and Europeans arrived in Russia's far Northeast during the early 17th century, the Whale Alley culture began a steep decline. One authority says that unfavorable ecological conditions from the 15th through the 18th centuries changed whale migration patterns and made Whale Alley untenable. This, combined with an element of cultural shock, may have led the Masigmit to reshape their social and spiritual institutions, abandon their whaling lifestyle and leave behind their eerie cities of bone. Where they went and what happened to them remain a mystery.

Today, whales are safe in Whale Alley. But Native hunters still kill Pacific walruses from a population of about 20,000 that use this area.

From the eastern tip of Arakamchechen, where Alaska is only 200 miles to the east, a 100-foot seaside cliff overlooks a walrus "ugli." It is a kind of walrus riviera where the boys haul out to rest during summer. (Females with their calves go much farther north to the safety of the permanent ice pack.) The site, one of 11 uglis along this coast, has been designated a protected walrus refuge.

About 2,000 walrus bulls lie stretched out here, looking like huge Vienna sausage hors d'oeuvres, complete with toothpicks. Most are asleep, but individuals constantly push their way in or force their way out of the prostrate herd. A 3,000-pound walrus steps on the flipper of a sleeping neighbor that bellows in rage and rolls over another that stabs his tusk into the back of a third that rears up and threatens a fourth until half a dozen bulls are waving their tusks at each other and creating a fearful din.

Several species of gulls are on hand to pick up small marine animals driven to the surface by the walrus traffic. Among them is a slaty-backed gull, especially attractive to birders who maintain life lists. The equally rare common ringed plover and bean goose are here too. On this day, 41 species, most of them sea- and shorebirds, prowl the water's edge for food.

Around the corner in Whale Alley, 20 gray whales, including a female with a calf, are scooping amphipods, tiny shrimplike creatures, from the bottom. A feeding whale dives, plows the bottom for a hundred yards, then surfaces with a mouthful of brown volcanic silt. It sifts out the amphipods in a whorl of tawny water. Blowing a 30-foot plume and taking a deep breath, it raises its flukes and dives for the bottom again.

In late August, nights grow longer, but we still have 18 hours of daylight. Sandhill cranes are our reliable alarm clocks. They pass over our camp at first light while bugling to each other, sounds evocative of the dawn of time.

Behind our tent camp lies a broad talus slope. Its boulders shelter families of lemming, arctic ground squirrel and short-tailed weasel (ermine). They show little fear and often scamper through camp. A weasel puts on a show for us,

Seven Steller sea lion skulls call living sea lions to slaughter at this Whale Alley hunting site. Local people of the Chukchi Peninsula avow no relation to the Whale Alley people whose culture disappeared sometime after the early 17th century. (Downs Matthews)

LEFT: *Male walrus haul out on Arakamchechen Island while females and young swim far northward to the safety of the pack ice. (Downs Matthews)*

FACING PAGE: *Yupik commercial walrus hunters from the village of New Chaplino butcher their catch off Yttygran Island. The hunters are employed by the state to hunt walrus to provide meat for silver fox farms, the nearest one being at Yanrakynnot. The villagers raise blue phase arctic foxes for their pelts, feeding the animals walrus meat. The tusks go to the state ivory factory to be carved into tourist items, the teeth are sold as souvenirs or kept by the hunters who make totems. At last report, this commerce was ongoing. (Downs Matthews)*

racing back and forth, pirouetting like a ballet dancer and bouncing up and down as if on a tundra trampoline.

On a flight inland, we see two silver-backed male grizzlies. One gallops off and hides. The other simply runs in circles. Later we see a female grizzly with three cubs and land on a flower-decked hilltop to take pictures. She ducks into a narrow swale and vanishes. Probably a good thing,

To photograph common and thick-billed murres, black-legged kittiwakes and horned puffins, we board two hand-built walrus-skin boats manned by Yupik Eskimos from Chaplino and motor out to Small Island. We pass through flocks of least auklets and crested auklets, murrelets and pelagic cormorants resting on the water's surface. At the cliffs, we land on a tiny beach to photograph kittiwake chicks.

On another day, we land our helicopter near three boatloads of Chaplino men who have killed and are butchering a walrus. It is a gory business, but the experience affords a rare opportunity to learn about subsistence hunting.

Our Eskimo guides accept a present of walrus meat for dinner and offer to share. The meat tastes a bit like coarse beef marinated in cod liver oil. Walrus *manguna*, the equivalent of whale muktuk, looks and eats like crunchy cork with a fishy flavor. It's an acquired taste, but one you must get used to if you live in Whale Alley.

EDITOR'S NOTE: *Nationally known writer Downs Matthews has written several articles for* ALASKA GEOGRAPHIC®. *He visited Chukotka in the early 1990s where he encountered the Whale Alley culture.*

Bibliography

Anderson, Paul J. and John F. Piatt. "Trophic Reorganization In The Gulf of Alaska Following Ocean Climate Regime Shift," in *Marine Ecology, Progress Series*. In press.

Bennis, Francine, Dorothy Childers and Steve Ganey. *Bycatch: Wasting Alaska's Future*. Anchorage: Alaska Marine Conservation Council, 1998.

Francis, Robert C., Steven R. Hare, Anne B. Hollowed and Warren S. Wooster. "Effects of interdecadal climate variability on the oceanic ecosystems of the NE Pacific," in *Fisheries Oceanography*, 7:1, 1-21, 1998.

Gay, Joel. *Commercial Fishing in Alaska*. Anchorage: Alaska Geographic Society, 1997.

Gibson, Margie Ann and Sallie B. Schullinger. *Answers From the Ice Edge*. Anchorage: Arctic Network and Greenpeace USA, 1998.

Macklin, S. Allen ed. *Bering Sea FOCI (Fisheries-Oceanography Coordinated Investigations) Final Report, 1991-1997*. NOAA/ Pacific Marine Environmental Laboratory contribution 1964; Fisheries-Oceanography Coordinated Investigations contribution B356. Springfield, Va.: National Technical Information Service, Dec. 1998.

MacLeish, Sumner. *Seven Words for Wind*. Seattle: Epicenter Press, 1997.

National Research Council. *The Bering Sea Ecosystem*. Washington, D.C.: National Academy Press, 1996.

Rennick, Penny ed. *Alaska's Seward Peninsula*. Anchorage: Alaska Geographic Society, 1987.

Schumacher, James D. and Phyllis J. Stabeno. "Continental Shelf of the Bering Sea," ch. 27, pp. 789-822 in *The Sea: The Global Coastal Ocean: Regional Studies and Syntheses*, Vol. 11. Allan R. Robinson and Kenneth H. Brink, eds. New York: John Wiley and Sons, 1998.

Index

PHOTOGRAPHERS

ALASKA GEOGRAPHIC. Back Issues

The North Slope, Vol. 1, No. 1. Out of print.
One Man's Wilderness, Vol. 1, No. 2. Out of print.
Admiralty...Island in Contention, Vol. 1, No. 3. $19.95.
Fisheries of the North Pacific, Vol. 1, No. 4. Out of print.
Alaska-Yukon Wild Flowers, Vol. 2, No. 1. Out of print.
Richard Harrington's Yukon, Vol. 2, No. 2. Out of print.
Prince William Sound, Vol. 2, No. 3. Out of print.
Yakutat: The Turbulent Crescent, Vol. 2, No. 4. Out of print.
Glacier Bay: Old Ice, New Land, Vol. 3, No. 1. Out of print.
The Land: Eye of the Storm, Vol. 3, No. 2. Out of print.
Richard Harrington's Antarctic, Vol. 3, No. 3. $19.95.
The Silver Years, Vol. 3, No. 4. $19.95.
Alaska's Volcanoes, Vol. 4, No. 1. Out of print.
The Brooks Range, Vol. 4, No. 2. Out of print.
Kodiak: Island of Change, Vol. 4, No. 3. Out of print.
Wilderness Proposals, Vol. 4, No. 4. Out of print.
Cook Inlet Country, Vol. 5, No. 1. Out of print.
Southeast: Alaska's Panhandle, Vol. 5, No. 2. Out of print.
Bristol Bay Basin, Vol. 5, No. 3. Out of print.
Alaska Whales and Whaling, Vol. 5, No. 4. $19.95.
Yukon-Kuskokwim Delta, Vol. 6, No. 1. Out of print.
Aurora Borealis, Vol. 6, No. 2. $19.95.
Alaska's Native People, Vol. 6, No. 3. Out of print.
The Stikine River, Vol. 6, No. 4. $19.95.
Alaska's Great Interior, Vol. 7, No. 1. $19.95.
Photographic Geography of Alaska, Vol. 7, No. 2. Limited.
The Aleutians, Vol. 7, No. 3. Out of print.
Klondike Lost, Vol. 7, No. 4. Out of print.
Wrangell-Saint Elias, Vol. 8, No. 1. Limited.
Alaska Mammals, Vol. 8, No. 2. Out of print.
The Kotzebue Basin, Vol. 8, No. 3. Out of print.

ALASKA GEOGRAPHIC. Notecard Set

$12.95

(+ $3.50 shipping/handling)

Each set includes 12 blank cards with envelopes, three cards each of four popular images from the cover of *ALASKA GEOGRAPHIC®*. Order from:

Alaska Geographic Society
P.O. Box 93370, Anchorage AK 99509-3370
907-562-0164; fax 907-562-0479; e-mail info@akgeo.com

Alaska National Interest Lands, Vol. 8, No. 4. $19.95.
Alaska's Glaciers, Vol. 9, No. 1. Revised 1993. $19.95.
Sitka and Its Ocean/Island World, Vol. 9, No. 2. Out of print.
Islands of the Seals: The Pribilofs, Vol. 9, No. 3. $19.95.
Alaska's Oil/Gas & Minerals Industry, Vol. 9, No. 4. $19.95.
Adventure Roads North, Vol. 10, No. 1. $19.95.
Anchorage and the Cook Inlet Basin, Vol. 10, No. 2. $19.95.
Alaska's Salmon Fisheries, Vol. 10, No. 3. $19.95.
Up the Koyukuk, Vol. 10, No. 4. $19.95.
Nome: City of the Golden Beaches, Vol. 11, No. 1. $19.95.
Alaska's Farms and Gardens, Vol. 11, No. 2. $19.95.
Chilkat River Valley, Vol. 11, No. 3. $19.95.
Alaska Steam, Vol. 11, No. 4. $19.95.
Northwest Territories, Vol. 12, No. 1. $19.95.
Alaska's Forest Resources, Vol. 12, No. 2. $19.95.
Alaska Native Arts and Crafts, Vol. 12, No. 3. $24.95.
Our Arctic Year, Vol. 12, No. 4. $19.95.
Where Mountains Meet the Sea, Vol. 13, No. 1. $19.95.
Backcountry Alaska, Vol. 13, No. 2. $19.95.
British Columbia's Coast, Vol. 13, No. 3. $19.95.
Lake Clark/Lake Iliamna, Vol. 13, No. 4. Out of print.
Dogs of the North, Vol. 14, No. 1. $21.95.
South/Southeast Alaska, Vol. 14, No. 2. Limited.
Alaska's Seward Peninsula, Vol. 14, No. 3. $19.95.
The Upper Yukon Basin, Vol. 14, No. 4. $19.95.
Glacier Bay: Icy Wilderness, Vol. 15, No. 1. Limited.
Dawson City, Vol. 15, No. 2. $19.95.
Denali, Vol. 15, No. 3. $19.95.
The Kuskokwim River, Vol. 15, No. 4. $19.95.
Katmai Country, Vol. 16, No. 1. $19.95.
North Slope Now, Vol. 16, No. 2. $19.95.
The Tanana Basin, Vol. 16, No. 3. $19.95.
The Copper Trail, Vol. 16, No. 4. $19.95.
The Nushagak Basin, Vol. 17, No. 1. $19.95.
Juneau, Vol. 17, No. 2. Limited.
The Middle Yukon River, Vol. 17, No. 3. $19.95.
The Lower Yukon River, Vol. 17, No. 4. $19.95.
Alaska's Weather, Vol. 18, No. 1. $19.95.
Alaska's Volcanoes, Vol. 18, No. 2. $19.95.
Admiralty Island: Fortress of Bears, Vol. 18, No. 3. $21.95.
Unalaska/Dutch Harbor, Vol. 18, No. 4. $19.95.
Skagway: A Legacy of Gold, Vol. 19, No. 1. $19.95.
Alaska: The Great Land, Vol. 19, No. 2. $19.95.
Kodiak, Vol. 19, No. 3. Out of print.
Alaska's Railroads, Vol. 19, No. 4. $19.95.
Prince William Sound, Vol. 20, No. 1. $19.95.
Southeast Alaska, Vol. 20, No. 2. $19.95.
Arctic National Wildlife Refuge, Vol. 20, No. 3. $19.95.
Alaska's Bears, Vol. 20, No. 4. $19.95.
The Alaska Peninsula, Vol. 21, No. 1. $19.95.
The Kenai Peninsula, Vol. 21, No. 2. $19.95.
People of Alaska, Vol. 21, No. 3. $19.95.
Prehistoric Alaska, Vol. 21, No. 4. $19.95.
Fairbanks, Vol. 22, No. 1. $19.95.
The Aleutian Islands, Vol. 22, No. 2. $19.95.
Rich Earth: Alaska's Mineral Industry, Vol. 22, No. 3. $19.95.
World War II in Alaska, Vol. 22, No. 4. $19.95.
Anchorage, Vol. 23, No. 1. $21.95.
Native Cultures in Alaska, Vol. 23, No. 2. $19.95.
The Brooks Range, Vol. 23, No. 3. $19.95.
Moose, Caribou and Muskox, Vol. 23, No. 4. $19.95.
Alaska's Southern Panhandle, Vol. 24, No. 1. $19.95.
The Golden Gamble, Vol. 24, No. 2. $19.95.
Commercial Fishing in Alaska, Vol. 24, No. 3. $19.95.
Alaska's Magnificent Eagles, Vol. 24, No. 4. $19.95.
Steve McCutcheon's Alaska, Vol. 25, No. 1. $21.95.
Yukon Territory, Vol. 25, No. 2. $21.95.
Climbing Alaska, Vol. 25, No. 3. $21.95.
Frontier Flight, Vol. 25, No. 4. $21.95. Our 100th Issue!
Restoring Alaska: Legacy of an Oil Spill, Vol. 26, No. 1. $21.95.
World Heritage Wilderness, Vol. 26, No. 2. $21.95.

PRICES AND AVAILABILITY SUBJECT TO CHANGE

Membership in The Alaska Geographic Society includes a subscription to *ALASKA GEOGRAPHIC®*, the Society's colorful, award-winning quarterly.

Call or write for current membership rates or to request a free catalog. *ALASKA GEOGRAPHIC®* back issues are also available (see above list). **NOTE:** This list was current in late-1999. If more than a year or two has elapsed since that time, contact us before ordering to check prices and availability of back issues, particularly books marked Limited.

When ordering back issues please add $4 for the first book and $2 for each additional book ordered for Priority Mail. Inquire for non-U.S. postage rates. To order, send check or money order (U.S. funds) or VISA/MasterCard information (including expiration date and your phone number) with list of titles desired to:

ALASKA GEOGRAPHIC.
P.O. Box 93370 • Anchorage, AK 99509-3370
Phone: (907) 562-0164 • Fax (907) 562-0479
Toll free (888) 255-6697 • E-mail: info@akgeo.com

NEXT ISSUE: **VOL. 26, NO. 4**

Russian America

From Kotzebue to Kodiak, from the Fox Islands of the Aleutians to Sitka, Alaska's Russian past lives on in place names, churches and people. Featuring archival and contemporary images, *Russian America* tells the story of Alaska under the czar, from earliest explorers who glimpsed both Asia and America from Bering Strait, to freebooting fur hunters, to the Alaska Purchase of 1867, which made the colony part of the United States for 2 cents an acre. To members winter 1999.